The Baofeng Radio Guide for Emergency Preparedness

Guerrilla Radio Strategies For Preppers and Survivalists

Rico Nery

Copyright © 2024 by Rico Nery

All rights reserved. No part of this publication may be reproduced, distributed, or transmitted in any form or by any means, including photocopying, recording, or other electronic or mechanical methods, without the prior written permission of the publisher.

CONTENTS

INTRODUCTION..1
    Purpose of the Guide... 1
    Who This Guide is For... 5
    How to Use This Guide.. 8

CHAPTER 1: Understanding Baofeng Radios.................12
    What is Baofeng Radio?... 12
    Types of Baofeng Radios... 17
    Basic Functions and Features.................................... 22

CHAPTER 2: Getting Started with Your Baofeng Radio 29
    Purchasing Your First Baofeng Radio....................... 29
    Unboxing and Setup.. 31
    Basic Operation... 33

CHAPTER 3: Essential Accessories for Your Baofeng Radio...37
    Antennas... 37
    Batteries.. 40
    Headsets and Microphones....................................... 43
    Other Accessories... 45

CHAPTER 4: Basic Radio Communication Principles....49
    Understanding Radio Waves.....................................49
    Frequency Bands.. 52
    Radio Etiquette and Protocols................................... 55

CHAPTER 5: Programming Your Baofeng Radio........... 60
    Manual Programming... 60
    Using Software for Programming..............................63
    Storing Frequencies.. 66

CHAPTER 6: Emergency Frequencies and Networks..... 70

  Emergency Channels.. 70
  Joining Local Radio Networks.. 72
  Monitoring Emergency Services...................................... 75
CHAPTER 7: Advanced Communication Techniques..... 80
  Repeater Use.. 80
  Signal Boosting..82
  Encryption and Privacy... 84
CHAPTER 8: Guerrilla Communication Strategies........ 88
  Introduction to Guerrilla Tactics......................................88
  Stealth Communication Techniques............................... 90
  Improvising Antennas and Power Sources..................... 92
CHAPTER 9: Setting Up a Base Station............................96
  Choosing a Location... 96
  Equipment and Setup..98
  Maintaining Your Base Station....................................... 103
CHAPTER 10: Mobile and Field Operations...................107
  Using Your Radio in the Field.. 107
  Vehicle Installation.. 110
  Field Antennas and Power Sources............................... 113
CHAPTER 11: Survival Communication Scenarios........ 118
  Natural Disasters...118
  Man-Made Disasters.. 122
  SHTF Scenarios.. 125
CHAPTER 12: Prepping Essentials................................. 131
  Building a Comprehensive Prepper's Kit......................132
  Creating a Family Communication Plan...................... 135
  Long-Term Survival Strategies.......................................138
CHAPTER 13: Troubleshooting and Maintenance......... 143

Common Issues and Solutions..143
Routine Maintenance...146
Repair and Spare Parts...148
Chapter 14: Legal and Ethical Considerations................ 152
Licensing Requirements.. 152
Ethical Use of Radios.. 153
Privacy and Security... 155
CONCLUSION..160
Summary of Key Points..160
Building a Communication Plan....................................164
Final Thoughts.. 165

INTRODUCTION

Effective communication is a critical component in emergency preparedness and survival. In times of crisis, whether natural disasters, man-made emergencies, or societal collapse, having a reliable means of communication can mean the difference between life and death. This guide, "The Baofeng Radio Guide for Emergency Preparedness: Guerrilla Radio Strategies For Preppers and Survivalists", aims to equip readers with the knowledge and skills necessary to effectively utilize Baofeng radios. Baofeng radios, renowned for their affordability and versatility, are an essential tool for anyone serious about emergency preparedness.

Purpose of the Guide

The primary goal of this guide is to provide comprehensive information on the use of Baofeng radios for emergency preparedness. While the focus is on the technical aspects of Baofeng radios, this guide also delves into broader strategies and techniques for preppers and survivalists. The intent is to not only familiarize readers with the operation and capabilities of Baofeng radios but also to integrate these radios into a holistic emergency communication plan. By doing so, individuals can ensure they are prepared to maintain contact with family, friends, and emergency services when conventional communication channels fail.

Importance of Communication in Emergencies

Communication is vital during emergencies for several reasons. First and foremost, it allows individuals to stay informed about the situation. Access to real-time information can guide decision-making, from evacuation routes to the availability of resources. Furthermore, communication enables coordination and support. Whether coordinating with family members, neighbors, or emergency responders, being able to relay and receive messages ensures a unified and effective response to the crisis.

In survival scenarios, isolation can lead to panic and poor decision-making. Having the means to communicate reduces anxiety and fosters a sense of connection and solidarity. Additionally, communication is essential for calling for help. In situations where individuals are trapped or injured, the ability to reach out to emergency services can expedite rescue and provide life-saving assistance.

Overview of Baofeng Radios and Their Applications

Baofeng radios have gained popularity among preppers and survivalists due to their cost-effectiveness, durability, and versatility. These radios offer a range of features that make them suitable for various communication needs in emergency situations.

Baofeng radios are handheld transceivers that operate on both Very High Frequency (VHF) and Ultra High Frequency (UHF) bands. This dual-band capability allows users to access a wide range of frequencies, making it easier to find clear channels for communication.

Additionally, many Baofeng models come equipped with features such as programmable channels, dual watch functionality, and emergency alarms, enhancing their utility in critical situations. One of the key advantages of Baofeng radios is their programmability. Users can manually program frequencies or use software to configure the radios for specific needs. This flexibility is particularly useful for preppers who may need to communicate on various local, regional, or national frequencies depending on the nature of the emergency. Baofeng radios also support the use of repeaters, which can significantly extend the communication range. In areas where direct line-of-sight communication is challenging due to terrain or obstructions, repeaters help bridge the gap, ensuring messages can travel over greater distances. This feature is invaluable in both urban and rural settings, where maintaining communication across large areas is crucial.

Baofeng radios are known for their ruggedness. Built to withstand harsh conditions, these radios are reliable in environments where more delicate equipment might fail. Whether dealing with extreme weather, rough handling, or prolonged use, Baofeng radios are designed to endure and continue functioning.

Beyond emergency preparedness, Baofeng radios have applications in various scenarios such as hiking, camping, and outdoor adventures. Their compact size and portability make them ideal for on-the-go communication, ensuring that users can stay connected even in remote locations. For survivalists, the ability to communicate while on the move is essential, as it allows for coordination, navigation, and safety in unfamiliar or hostile environments.

Who This Guide is For

The Baofeng Radio Guide for Emergency Preparedness: Guerrilla Radio Strategies for Preppers and Survivalists* is meticulously crafted for individuals who prioritize preparedness and self-reliance in times of crisis. **This guide is particularly valuable for**

preppers, survivalists, and those with a keen interest in guerrilla communication strategies.

Preppers are individuals or groups who actively prepare for emergencies, disruptions, or disasters. This preparation often includes stockpiling supplies, developing survival skills, and establishing communication plans. In the prepper community, maintaining communication is paramount. Whether dealing with natural disasters like hurricanes, earthquakes, or wildfires, or preparing for man-made crises such as power grid failures or economic collapse, preppers understand that staying informed and connected can significantly enhance their chances of survival. This guide provides preppers with detailed instructions on using Baofeng radios, ensuring they can maintain communication with family, friends, and other community members during emergencies.

Survivalists share similarities with preppers but often focus on developing a broader range of skills to thrive in various scenarios, from wilderness survival to urban escape strategies. Survivalists prioritize adaptability and resourcefulness, and effective communication is a cornerstone of their preparedness plans. Baofeng radios, with their versatility and durability, are ideal tools for survivalists who need reliable communication in diverse environments. This guide delves into advanced techniques and field operations, enabling survivalists to leverage their Baofeng radios to their fullest potential.

For those interested in guerrilla communication strategies, this guide offers valuable insights into stealthy and secure communication methods. Guerrilla communication involves techniques used by small, mobile groups to communicate covertly, often in hostile environments. This includes avoiding detection by adversaries, using improvised equipment, and maintaining operational security. This guide provides a thorough exploration of these strategies, integrating them with the technical capabilities of Baofeng radios to create a robust communication plan.

How to Use This Guide

This guide is designed to be comprehensive and user-friendly, providing readers with a structured approach to mastering Baofeng radios and integrating them into their overall emergency preparedness plans. To maximize the benefits of this guide, readers are encouraged to approach it methodically, progressing through each section and applying the knowledge and skills learned.

Begin by reading the introductory chapters to gain a solid understanding of the importance of communication in emergencies and the specific features and benefits of Baofeng radios. These foundational chapters set the stage for more detailed and technical information to come. Familiarize yourself with the basic operations and functions of your Baofeng radio before delving into advanced topics.

The chapters are organized to build on each other, starting with basic principles and gradually introducing more complex concepts and techniques. For those new to Baofeng radios, the initial chapters on understanding the radios and getting started will provide essential knowledge. These sections cover topics such as purchasing, unboxing, setup, and basic operation. Readers will learn how to turn on the radio, navigate basic settings, and perform their first transmission.

As you gain confidence and familiarity with your Baofeng radio, proceed to the chapters on essential accessories, programming, and advanced communication techniques. These sections will enhance your radio's functionality and improve your communication capabilities. Detailed instructions on manual programming, using software tools, and storing frequencies are provided to ensure you can tailor your radio to your specific needs. Additionally, chapters on emergency frequencies, repeater use, and signal boosting will help you extend your communication range and improve clarity.

The guide also includes chapters dedicated to guerrilla communication strategies and field operations. These sections are particularly valuable for survivalists and those interested in covert communication methods. Learn about stealth communication techniques, improvising antennas and power sources, and maintaining operational security. Practical advice on setting up a base station, using your radio in the field, and vehicle installation is also provided.

For preppers, the chapters on survival communication scenarios and prepping essentials offer targeted advice on integrating Baofeng radios into broader preparedness plans. These sections cover various emergency scenarios, from natural disasters to societal collapse, and provide strategies for building a comprehensive prepper's kit and creating a family communication plan.

The troubleshooting and maintenance chapter is essential for ensuring your Baofeng radio remains in top condition. Learn how to diagnose and resolve common issues, perform routine maintenance, and make necessary repairs. This knowledge will help you maintain your radio's reliability and longevity.

The final chapters on legal and ethical considerations emphasize the importance of responsible and lawful radio use. Understand licensing requirements, ethical communication practices, and methods for protecting your privacy and security.

By following this guide, readers will develop a thorough understanding of how to effectively use Baofeng radios for emergency communication. Whether you are a prepper, survivalist, or interested in guerrilla communication strategies, this guide provides the knowledge and skills necessary to stay informed and connected.

CHAPTER 1: Understanding Baofeng Radios

What is Baofeng Radio?

Baofeng radios are handheld transceivers that have garnered a reputation for their affordability, versatility, and durability. These radios operate on both Very High Frequency (VHF) and Ultra High Frequency (UHF) bands, allowing them to communicate effectively over different distances and through various types of terrain. The

dual-band functionality is one of the key features that make Baofeng radios a popular choice among preppers, survivalists, amateur radio enthusiasts, and outdoor adventurers.
VHF frequencies, which range from 30 MHz to 300 MHz, are suitable for long-distance communication in open areas with minimal obstructions. This makes them ideal for rural settings, maritime communication, and scenarios where the line of sight is relatively unobstructed. UHF frequencies, ranging from 300 MHz to 3 GHz, are more effective in urban environments and areas with dense vegetation because they can better penetrate buildings and other obstacles. This flexibility in frequency usage ensures that Baofeng radios can perform reliably in diverse settings, enhancing their appeal for emergency preparedness and survival scenarios.

A distinguishing feature of Baofeng radios is their programmability. Users can manually program frequencies or utilize software tools to input and organize channels according to their specific communication needs. This level of customization allows for efficient communication in various situations, whether coordinating with local emergency services, communicating within a community network, or keeping in touch with family members.

Baofeng radios are also equipped with a range of additional features designed to enhance their functionality in emergency situations. Many models include an emergency alarm function that can send a distress signal, dual watch functionality to monitor two frequencies simultaneously, and built-in flashlights for utility during power outages or nighttime operations. These features, combined with their robust construction, make Baofeng radios a reliable choice for those who require dependable communication tools.

History and Evolution

The history of Baofeng radios is intertwined with the broader development of radio communication technology. Radio communication has evolved significantly over the past century, with substantial advancements in both technology and accessibility.

The origins of radio communication date back to the late 19th and early 20th centuries when pioneers like Guglielmo Marconi and Nikola Tesla made groundbreaking discoveries in wireless transmission. Early radio equipment was large, expensive, and primarily used by the military and commercial entities. The advent of transistor technology in the 1950s revolutionized the industry by enabling the production of smaller, more affordable, and more efficient radios. This development democratized radio communication, making it accessible to a broader audience.

Baofeng, a Chinese electronics company, was founded in the early 2000s with the goal of producing high-quality radios at competitive prices. The company's mission was to create affordable communication tools that could meet the needs of a global market. Baofeng quickly established a reputation for manufacturing reliable and feature-rich radios that were significantly cheaper than those offered by established brands.

The release of the Baofeng UV-5R in 2012 marked a significant milestone in the company's history. The UV-5R was an instant success due to its affordability, versatility, and extensive range of features. Its dual-band capability, allowing operation on both VHF and UHF frequencies, made it a popular choice among amateur radio enthusiasts, preppers, and survivalists. The UV-5R's programmability and extensive feature set provided users with a powerful tool for a variety of communication needs.

Since the introduction of the UV-5R, Baofeng has continued to innovate and expand its product line. The company has introduced new models with enhanced features and improved performance, catering to the evolving needs of its diverse user base. Baofeng's commitment to quality and affordability has enabled it to become a dominant player in the global radio market, with millions of units sold worldwide.

Types of Baofeng Radios

Baofeng offers a variety of radio models, each designed to meet different communication needs. The range of Baofeng radios includes basic models suitable for beginners as well as advanced models with features tailored for experienced users.

The Baofeng UV-5R series is one of the most popular and widely recognized lines. The UV-5R is known for its dual-band capability, allowing operation on both VHF and UHF frequencies. It features a programmable memory, dual watch functionality, and a range of other useful features such as an emergency alarm, built-in flashlight, and FM radio receiver. The UV-5R is highly customizable, with support for various accessories and enhancements.

Another popular model is the Baofeng BF-F8HP, which is an upgraded version of the UV-5R. The BF-F8HP offers higher output power, which extends its communication range and improves signal clarity. It also features a larger battery for longer operational life and a more robust construction for enhanced durability. The BF-F8HP is ideal for users who require more power and durability in their communication tools.

The Baofeng UV-82 series is another notable line, known for its improved ergonomics and enhanced features. The UV-82 radios have a larger, more comfortable keypad and a more powerful speaker, providing better audio quality. They also feature dual push-to-talk buttons, allowing users to switch between two frequencies without changing channels. This makes the UV-82 series ideal for situations where quick and efficient communication is crucial.

For users who require advanced features and capabilities, the Baofeng UV-9R Plus is a robust choice. The UV-9R Plus is designed for rugged environments, with waterproof and dustproof construction. It offers higher power output, extended battery life, and enhanced audio quality. The UV-9R Plus is suitable for outdoor adventures, emergency response, and other demanding applications.

Popular Models and Their Features

Baofeng radios come in various models, each with unique features and capabilities. Here are some of the most popular models and their key features:

Baofeng UV-5R

The UV-5R is the flagship model of Baofeng radios and is known for its versatility and affordability. **Key features include:**

- Dual-band operation (VHF/UHF)
- Programmable memory for storing frequencies
- Dual watch functionality for monitoring two frequencies simultaneously
- Built-in flashlight for utility in low-light conditions
- Emergency alarm function for sending distress signals
- FM radio receiver for listening to local broadcasts
- Support for various accessories, including extended batteries and external antennas

Baofeng BF-F8HP

The BF-F8HP is an upgraded version of the UV-5R, offering enhanced performance and features. Key features include:
- Higher output power (up to 8 watts) for extended communication range
- Larger battery for longer operational life
- Improved construction for better durability

- ❖ Enhanced audio quality for clearer communication
- ❖ Dual-band operation (VHF/UHF) and programmable memory
- ❖ Compatibility with UV-5R accessories

Baofeng UV-82

The UV-82 series is known for its ergonomic design and enhanced audio features. Key features include:

- ❖ Larger, more comfortable keypad for easier operation
- ❖ Dual push-to-talk buttons for quick switching between frequencies
- ❖ More powerful speaker for better audio quality
- ❖ Dual-band operation (VHF/UHF) and programmable memory
- ❖ Built-in flashlight and emergency alarm function
- ❖ Support for a wide range of accessories

Baofeng UV-9R Plus

The UV-9R Plus is designed for rugged environments and demanding applications. Key features include:
- ❖ Waterproof and dustproof construction for durability in harsh conditions
- ❖ Higher output power (up to 8 watts) for extended communication range
- ❖ Extended battery life for prolonged use in the field
- ❖ Enhanced audio quality for clearer communication
- ❖ Dual-band operation (VHF/UHF) and programmable memory
- ❖ Built-in flashlight and emergency alarm function
- ❖ Ideal for outdoor adventures, emergency response, and other demanding scenarios

Basic Functions and Features

Baofeng radios are equipped with a range of basic functions and features that enhance their utility and versatility. Understanding these functions and features is essential for effectively using your Baofeng radio.

Keypad and Display

The keypad and display are critical components of Baofeng radios. The keypad is used to input frequencies, navigate menus, and access various functions. Most Baofeng radios feature a backlit keypad, allowing for easy operation in low-light conditions. The display shows important information such as the current frequency, battery level, signal strength, and menu options. It is typically an LCD screen that is easy to read and provides clear information.

Battery

Baofeng radios are powered by rechargeable lithium-ion batteries, which provide long-lasting power for extended use. The batteries are designed to be easily replaceable, allowing users to carry spare batteries for prolonged operations. Charging the batteries is straightforward, with most models including a desktop charger or USB charging option. The battery life varies depending on the model and usage, but Baofeng radios are generally known for their efficient power consumption and long operational life.

Antenna

The antenna is a crucial component of any radio, as it determines the range and quality of communication. Baofeng radios come with detachable antennas, allowing users to upgrade to higher-gain antennas for extended range and better signal clarity. The standard antennas provided with Baofeng radios are sufficient for most general use cases, but for more demanding applications, upgrading to a higher-gain antenna can significantly enhance performance.

Programmability

One of the standout features of Baofeng radios is their programmability. Users can manually program frequencies directly on the radio or use software tools such as CHIRP to input and organize channels. This programmability allows for extensive customization based on specific communication needs. For example, users can program their radios with local emergency frequencies, community channels, and personal communication frequencies. This level of customization ensures that users have access to all necessary communication channels during emergencies.

Dual-Band Operation
Baofeng radios are capable of dual-band operation, meaning they can operate on both VHF and U
HF frequencies. This flexibility allows users to switch between different frequency bands depending on the communication requirements and environmental conditions. VHF frequencies are ideal for long-distance communication in open areas, while UHF frequencies are better suited for urban environments and areas with dense vegetation. Dual-band operation ensures that Baofeng radios can perform reliably in a variety of settings.

Additional Features

Baofeng radios come with a range of additional features that enhance their utility. These features include:

Emergency Alarm: Many Baofeng models include an emergency alarm function that can be activated to send a distress signal. This feature is particularly useful in emergency situations where immediate assistance is needed.

Dual Watch: Dual watch functionality allows the radio to monitor two frequencies simultaneously. This ensures that users do not miss important communications while listening to another channel.

Built-in Flashlight: A built-in flashlight is a common feature in many Baofeng models. This provides additional utility during power outages or nighttime operations, making the radio a versatile tool in various situations.

FM Radio Receiver: Some Baofeng radios include an FM radio receiver, allowing users to listen to local broadcasts. This feature can be useful for staying informed about news and weather updates, especially during emergencies.

VOX (Voice Operated Exchange): VOX functionality enables hands-free operation by automatically transmitting when the user speaks into the microphone. This feature is useful in situations where hands-free communication is needed, such as when using the radio while driving or performing tasks.

Baofeng radios are versatile, reliable, and affordable communication tools that serve a wide range of needs. Their dual-band capability, programmability, and additional features make them ideal for emergency communication, recreational use, and outdoor activities. By understanding the history, technical features, and practical applications of Baofeng radios, users can maximize their utility and ensure they are prepared for any communication challenges they may face. This chapter has provided a comprehensive overview of what Baofeng radios are, their evolution, the different types and models available, and the basic functions and features that make them indispensable tools for preppers, survivalists, and anyone needing reliable communication. The following chapters will delve deeper into the practical usage, advanced programming techniques, and strategies for effective communication in various scenarios.

CHAPTER 2: Getting Started with Your Baofeng Radio

Purchasing Your First Baofeng Radio

When it comes to purchasing your first Baofeng radio, several factors need to be considered to ensure you select the right model for your needs. The vast array of models and features can be overwhelming, but understanding your specific requirements will help you make an informed decision.

First, consider the primary use of the radio. If you need a radio for emergency preparedness, look for features such as dual-band capability, programmability, and durability. For outdoor activities like hiking or camping, a rugged model with a long battery life and weather resistance may be ideal. If you are an amateur radio enthusiast, you might prioritize features like power output, advanced programming options, and accessory compatibility.

Next, evaluate the power output of the radio. Higher wattage generally means better range and signal clarity. Most Baofeng radios offer power settings ranging from 1 watt to 8 watts. For general use, a radio with 4 to 5 watts should suffice, but for more demanding applications, consider a model with higher power output.

Battery life is another critical factor. Look for radios with long-lasting batteries, especially if you plan to use the radio for extended periods without access to charging facilities. Models with larger battery capacities or the option to use AA or AAA batteries can provide additional flexibility.

Consider the accessories that come with the radio or can be purchased separately. Essential accessories include a programming cable, spare batteries, an external antenna, and a speaker-microphone. These accessories can enhance the functionality and convenience of your Baofeng radio.

Finally, ensure the radio is compliant with local regulations. In some regions, certain frequencies and power outputs may be restricted. Check with your local communications authority to understand any legal requirements or limitations.

Unboxing and Setup

Once you have purchased your Baofeng radio, the next step is to unbox and set it up. The unboxing process is straightforward, but it's essential to familiarize yourself with the components and accessories included in the package.

When you open the box, you will typically find the following items: the Baofeng radio, a rechargeable battery, a charging dock or cable, a belt clip, a wrist strap, an antenna, a user manual, and sometimes an earpiece or speaker-microphone. Carefully remove each item from the box and ensure nothing is missing.

Begin by attaching the antenna to the radio. Screw the antenna onto the SMA connector on the top of the radio until it is securely fastened.

The antenna is crucial for transmitting and receiving signals, so ensure it is properly attached.

Next, insert the battery into the radio. Align the battery with the contacts on the back of the radio and slide it into place until it clicks. Make sure the battery is fully charged before using the radio for the first time. If the battery is not charged, place the radio in the charging dock or connect it to the charger and allow it to charge fully.

Attach the belt clip and wrist strap if desired. The belt clip allows you to securely attach the radio to your belt or bag, making it easily accessible during use. The wrist strap provides additional security, preventing accidental drops.

Once the radio is assembled and the battery is charged, it is time to turn on the radio and begin the initial setup.

Basic Operation

Turning on your Baofeng radio for the first time is an exciting moment. The initial setup and basic operation are simple, but it's essential to understand the key functions and settings to get the most out of your radio.

To turn on the radio, locate the volume knob on the top of the unit. Turn the knob clockwise until you hear a click and see the display light up. The radio will power on, and you will hear a brief startup sound.

The display will show the current frequency or channel, battery level, and other status indicators. Take a moment to familiarize yourself with the display and the various symbols and numbers it shows.

The keypad on the front of the radio is used to input frequencies, navigate menus, and adjust settings. The buttons are typically labeled with numbers, letters, and function symbols. The "Menu" button allows you to access the radio's menu system, where you can configure various settings and options.

To begin, set the radio to a frequency or channel that you want to use for communication. If you know the specific frequency, you can enter it directly using the keypad. Press the "VFO/MR" button to switch to frequency mode, then use the keypad to input the desired frequency. Press the "Menu" button to save the frequency.

For channel-based operation, press the "VFO/MR" button to switch to memory mode. Use the arrow keys to scroll through the pre-programmed channels. You can program channels with specific frequencies, names, and settings using a computer and programming software, which will be covered in a later chapter.

Once you have selected a frequency or channel, it's time to make your first transmission. Hold down the push-to-talk (PTT) button on the side of the radio while speaking clearly into the microphone. Release the PTT button when you finish speaking to listen for a response. Ensure you are following proper radio etiquette and any local regulations regarding radio communication.

Adjust the volume to a comfortable level using the volume knob. You can also use the keypad to access additional functions, such as activating the built-in flashlight, setting the squelch level, or adjusting the power output.

Familiarize yourself with other essential functions of your Baofeng radio. These may include scanning for active frequencies, setting up a dual watch to monitor two frequencies simultaneously, and using the radio's built-in FM receiver to listen to local broadcasts. Understanding these basic operations and functions will ensure you can effectively use your Baofeng radio for communication in various scenarios. As you become more comfortable with the radio, you can explore advanced features and programming options to further enhance its capabilities.

CHAPTER 3: Essential Accessories for Your Baofeng Radio

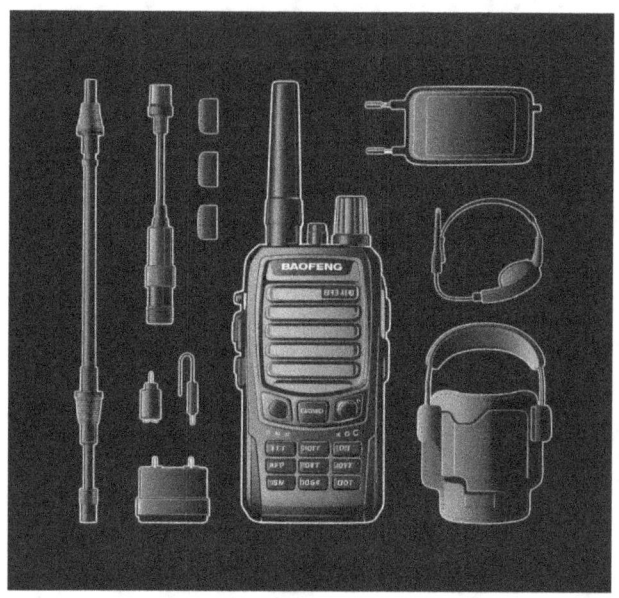

Antennas

Antennas are one of the most critical accessories for your Baofeng radio. The performance of your radio is heavily dependent on the quality and type of antenna you use. While the standard antenna that comes with your Baofeng radio is adequate for basic use, upgrading to a high-quality antenna can significantly enhance your radio's range and signal clarity.

There are several types of antennas available for Baofeng radios, each with distinct advantages. The most common types are whip antennas, rubber duck antennas, and telescopic antennas. **Whip antennas** are flexible and longer than standard antennas, usually made from a combination of metal and rubber materials. Their increased length allows for better signal reception and transmission, making them ideal for use in areas with challenging terrain or in situations where extended range is crucial. A popular choice among Baofeng users is the Nagoya NA-771, known for its superior performance and durability.

Rubber duck antennas, often referred to as flexible antennas, are shorter and more compact. They are typically made of rubber or plastic and are designed to withstand physical wear and tear. While they may not offer the same range as whip antennas, their rugged construction makes them suitable for outdoor activities and environments where the antenna might be subjected to physical stress.

Telescopic antennas are adjustable in length, allowing users to extend or retract the antenna based on their needs. This versatility makes telescopic antennas a good option for those who require both portability and enhanced range. However, they can be more fragile than whip or rubber duck antennas, so careful handling is necessary.

Upgrading your antenna can provide several benefits. A better antenna can improve the overall performance of your Baofeng radio by enhancing its ability to receive weak signals and transmit over longer distances. This is particularly important in emergency situations where clear communication can be a matter of life and death. Additionally, an upgraded antenna can reduce interference and provide

more consistent signal quality, ensuring that your communications are reliable and clear.
When selecting an antenna, consider factors such as length, flexibility, durability, and compatibility with your specific Baofeng model. Investing in a high-quality antenna can significantly improve your radio's performance and ensure you are prepared for any communication challenges.

Batteries

Batteries are another essential accessory for your Baofeng radio. The ability to stay connected in an emergency or remote location often depends on having reliable and long-lasting power sources. Understanding the different types of batteries and their benefits can help you make the best choice for your needs.

Baofeng radios typically come with a standard rechargeable lithium-ion battery, which provides a balance of capacity, weight, and rechargeability. While these batteries are suitable for general use, having extended or backup batteries can be invaluable, especially in situations where access to power is limited. Extended batteries offer a higher capacity than standard batteries, allowing for longer operation between charges. For instance, a standard Baofeng battery might offer a capacity of around 1800 mAh, while extended batteries can provide capacities of 2800 mAh or more. This increased capacity is particularly useful for long outings, field operations, or emergency scenarios where recharging might not be immediately possible. Extended batteries are slightly larger and heavier, but the trade-off is usually worth it for the added operational time.

Backup batteries are essential for ensuring continuous operation of your radio. Keeping spare batteries on hand means you can quickly swap out a depleted battery for a fully charged one, minimizing downtime. It's a good practice to regularly cycle through your backup batteries to ensure they are always in good condition and ready to use when needed.

For maximum versatility, consider batteries that are compatible with multiple charging methods. Some Baofeng radios support USB charging, which can be convenient if you have access to power banks, solar chargers, or car chargers. This flexibility ensures that you can keep your radio powered in a variety of situations.

Maintaining your batteries properly can extend their lifespan and ensure reliable performance. Store batteries in a cool, dry place and avoid exposing them to extreme temperatures. Regularly check the battery contacts for dirt or corrosion and clean them as necessary. When not in use, it is advisable to keep batteries partially charged, as storing them at full charge or fully depleted can reduce their overall lifespan. Investing in high-quality, extended, and backup batteries will ensure that your Baofeng radio remains operational for extended periods, providing you with reliable communication when you need it most.

Headsets and Microphones

Improving communication clarity is crucial, especially in noisy environments or situations where discretion is necessary. Headsets and microphones are valuable accessories that can enhance the functionality and usability of your Baofeng radio.

Headsets come in various styles, including in-ear, over-ear, and earmuff designs. In-ear headsets are compact and lightweight, making them ideal for situations where portability and comfort are essential. They fit snugly in the ear, providing clear audio without being obtrusive. Over-ear headsets offer better sound isolation and can be more comfortable for extended use. They typically feature cushioned ear cups that cover the entire ear, blocking out ambient noise and providing clear communication. Earmuff designs are the most robust, offering the best noise isolation and comfort for long-duration use. They are commonly used in industrial, military, or emergency response scenarios where high levels of noise are present.

When selecting a headset, consider factors such as comfort, durability, and audio quality. Look for models with noise-canceling features, as these can significantly improve communication clarity by reducing background noise. Additionally, ensure the headset is compatible with your specific Baofeng model, as different models may have different connector types.

Microphones, including speaker-microphones and lapel microphones, can also enhance communication clarity and convenience. Speaker-microphones are handheld devices that combine a microphone and speaker into one unit. They are typically clipped to the user's clothing, allowing for hands-free communication. This setup is particularly useful in situations where quick and easy access to the radio is required, such as during tactical operations or while driving.

Lapel microphones, also known as lapel mics or clip-on mics, are small, discreet microphones that can be attached to the user's clothing near the collar or chest. They are ideal for situations where discreet communication is needed, such as in security or surveillance operations. Lapel microphones often come with a push-to-talk (PTT) button, allowing for easy activation of the radio without having to handle the radio itself. Both headsets and microphones can significantly improve the usability of your Baofeng radio by providing clear audio and hands-free operation. These accessories are particularly valuable in noisy environments, during physical activities, or in situations where quick and discreet communication is necessary.

Other Accessories

Added to antennas, batteries, headsets, and microphones, there are several other accessories that can enhance the functionality and convenience of your Baofeng radio. These accessories include cases, chargers, and other useful gear that can help you make the most of your radio in various situations.

Protective cases are essential for safeguarding your Baofeng radio from physical damage. These cases come in various forms, including hard-shell cases, soft pouches, and silicone covers. Hard-shell cases provide the highest level of protection, shielding your radio from impacts, drops, and harsh environmental conditions. Soft pouches offer moderate protection while being lightweight and easy to carry. Silicone covers provide basic protection against scratches and minor impacts, and they often improve grip, preventing accidental drops. When choosing a protective case, consider the level of protection you need based on your intended use. For outdoor adventures or rugged environments, a hard-shell case may be the best option. For everyday use or urban settings, a soft pouch or silicone cover may suffice.

Chargers are another crucial accessory for keeping your Baofeng radio powered and ready for use. In addition to the standard desktop charger that comes with most Baofeng radios, consider investing in additional charging options such as car chargers, USB chargers, and solar chargers. Car chargers are particularly useful for mobile operations, allowing you to charge your radio while driving. USB chargers provide flexibility, as they can be used with power banks, laptops, and other USB-compatible devices. Solar chargers are an excellent option for extended outdoor activities, providing a renewable power source when access to electricity is limited.

Another useful accessory is a programming cable, which allows you to connect your Baofeng radio to a computer for advanced programming and customization. Using software such as CHIRP, you can easily program frequencies, channels, and other settings, making it more convenient to set up and manage your radio. A programming cable can save time and effort, especially if you need to program multiple radios or frequently change settings.

Other accessories that can enhance your Baofeng radio experience include external speakers, signal amplifiers, and carrying straps. External speakers can improve audio quality and volume, making it easier to hear communications in noisy environments. Signal amplifiers, also known as repeaters, can extend the range of your radio by boosting the signal strength. Carrying straps and harnesses provide additional convenience, allowing you to carry your radio comfortably and access it quickly when needed.

Investing in these accessories can significantly enhance the functionality, convenience, and usability of your Baofeng radio. By choosing the right accessories for your needs, you can ensure that your radio is always ready to provide reliable communication in any situation.

CHAPTER 4: Basic Radio Communication Principles

Understanding Radio Waves

Radio waves are the foundation of wireless communication, enabling devices to transmit and receive information over distances without the need for physical connections. Understanding how radio waves work, their characteristics, and their limitations is essential for effective communication with your Baofeng radio.
Radio waves are a type of electromagnetic radiation, similar to visible light, but with much longer wavelengths. They are generated by a transmitter and propagate through the air at the speed of light. The basic properties of radio waves include frequency, wavelength, and amplitude, all of which influence how the waves behave and how they can be used for communication.

Frequency, measured in hertz (Hz), refers to the number of oscillations or cycles per second of the radio wave. The frequency of a radio wave determines its position within the electromagnetic spectrum and its corresponding wavelength. Lower frequencies have longer wavelengths, while higher frequencies have shorter wavelengths. For example, the VHF (Very High Frequency) band ranges from 30 MHz to 300 MHz, and the UHF (Ultra High Frequency) band ranges from 300 MHz to 3 GHz.

Wavelength is the distance between successive peaks of the wave and is inversely related to frequency. Longer wavelengths can travel further distances and penetrate obstacles better than shorter wavelengths. This is why lower frequency bands are often used for long-range communication, while higher frequency bands are better suited for short-range communication in urban environments.

Amplitude refers to the height of the wave, which determines the strength or power of the signal. Higher amplitude signals are stronger and can travel further, but they also require more power to transmit.

Radio waves can propagate in several ways, including ground wave, skywave, and line-of-sight propagation. Ground waves travel along the surface of the Earth and can cover long distances, especially at lower frequencies. Skywaves are reflected off the ionosphere, allowing them to travel beyond the horizon and cover very long distances. Line-of-sight propagation, common in VHF and UHF bands, requires a clear path between the transmitter and receiver, as obstacles can block or reflect the signal.

Despite their versatility, radio waves have limitations. They can be affected by various factors, such as terrain, weather, and interference from other signals. Physical obstacles like buildings, mountains, and dense vegetation can block or attenuate radio waves, reducing their effective range. Atmospheric conditions, such as rain, fog, and solar activity, can also impact signal strength and quality. Additionally, interference from other electronic devices or radio signals can cause noise and disrupt communication.

Understanding these basic principles of radio waves is crucial for optimizing the performance of your Baofeng radio and ensuring reliable communication in various environments.

Frequency Bands

Frequency bands are specific ranges of frequencies allocated for different types of communication. The most commonly used frequency bands for Baofeng radios are the VHF (Very High Frequency) and UHF (Ultra High Frequency) bands. Each band has distinct characteristics and uses, making them suitable for different applications.

The VHF band ranges from 30 MHz to 300 MHz and is often used for long-range communication. VHF signals have longer wavelengths, allowing them to travel greater distances and penetrate obstacles like trees and buildings more effectively than UHF signals. This makes VHF ideal for outdoor and rural communication, where there are fewer obstructions. Common applications of VHF include marine communication, aviation, and amateur radio.

Baofeng radios typically operate in the VHF range of 136 MHz to 174 MHz. Within this range, different frequencies are designated for various services and uses. For example, the 144 MHz to 148 MHz range is commonly used by amateur radio operators (also known as "hams"), while the 150 MHz to 174 MHz range is often allocated for public safety and commercial communication.

The UHF band ranges from 300 MHz to 3 GHz and is well-suited for short-range communication in urban environments. UHF signals have shorter wavelengths, which are better at penetrating buildings, walls, and other structures. This makes UHF ideal for indoor use and in densely populated areas where obstacles are more prevalent. Common applications of UHF include television broadcasting, land mobile radio systems, and wireless microphones.

Baofeng radios typically operate in the UHF range of 400 MHz to 520 MHz. Within this range, different frequencies are allocated for various services, such as business and industrial communication, public safety, and amateur radio. The 420 MHz to 450 MHz range is popular among amateur radio operators, while the 450 MHz to 470 MHz range is often used for commercial and public safety communication.

When selecting a frequency band for your Baofeng radio, consider the specific requirements of your communication needs. For long-range outdoor communication, VHF may be the better choice due to its superior propagation characteristics. For urban or indoor communication, UHF may be more effective due to its ability to penetrate obstacles.
It's important to note that different countries have different regulations regarding the use of radio frequencies. Always check with your local communications authority to ensure you are operating within the legal frequency ranges and adhering to any licensing requirements.

Radio Etiquette and Protocols

Proper radio etiquette and protocols are essential for effective and efficient communication. Following established guidelines ensures that messages are clear, concise, and understood by all parties involved. Whether you are using your Baofeng radio for emergency communication, recreational activities, or professional use, adhering to these principles will enhance your communication experience.

One of the fundamental principles of radio communication is clarity. Always speak clearly and at a moderate pace. Enunciate your words to ensure that your message is easily understood, even in noisy environments or under poor signal conditions. Avoid speaking too quickly or too slowly, as this can make it difficult for the listener to comprehend your message.
Keep your messages concise and to the point. Avoid unnecessary chatter or long-winded explanations. Use simple and direct language to convey your message effectively. This is particularly important in emergency situations,

where time is critical, and clear communication can save lives.

Before transmitting, always listen for ongoing communications on the frequency or channel. Interrupting others can cause confusion and disrupt important communications. If the channel is clear, proceed with your transmission. If it is busy, wait for a break in communication before transmitting.

When initiating a transmission, identify yourself and the intended recipient. This helps establish who is communicating and ensures that the message is directed to the correct person. For example, you might say, "Base station, this is Alpha 1," to indicate that Alpha 1 is calling the base station.

Use standard radio terminology and procedures to ensure consistency and understanding. Common terms include "over" to indicate the end of your transmission and that you are expecting a response, "out" to indicate the end of your communication, and "copy" to acknowledge that you have received and understood the message. Other terms, such as "affirmative" for yes and "negative" for no, help avoid misunderstandings that can occur with simple yes or no answers.

In emergency situations, use the term "break" to interrupt ongoing communication with a critical message. If you have urgent information, say "break, break" to indicate that your message takes priority. Always follow emergency protocols and prioritize life-saving communication over routine traffic.

Avoid using slang, jargon, or code words that may not be understood by all users. Stick to standard terminology and plain language to ensure that everyone on the frequency can understand your message. This is especially important when communicating with emergency services or during coordinated operations. Maintain professionalism and respect at all times. Avoid using inappropriate language, making personal comments, or engaging in arguments over the radio. Remember that radio frequencies are public and can be monitored by others, including regulatory authorities.

Regularly practice your radio communication skills to become proficient and confident. Participate in drills, exercises, and training sessions to improve your ability to communicate effectively under various conditions. Familiarize yourself with the specific protocols and procedures used by your organization, community, or radio group.
By adhering to proper radio etiquette and protocols, you can ensure that your communications are clear, concise, and effective. This not only enhances your own communication experience but also contributes to the overall efficiency and safety of radio communication networks.

Understanding the basic principles of radio waves, the characteristics and uses of different frequency bands, and the importance of proper radio etiquette and protocols is essential for effective communication with your Baofeng radio. These foundational concepts provide the knowledge and skills necessary to use your radio confidently and competently in any situation. As you continue to explore and practice these principles, you will become a proficient and reliable radio operator, capable of navigating the complexities of radio communication with ease.

CHAPTER 5: Programming Your Baofeng Radio

Manual Programming

Programming your Baofeng radio manually can seem daunting at first, but with a systematic approach, it becomes a manageable task. Manual programming involves entering frequencies, setting parameters, and saving them to memory channels directly from the radio's keypad. This skill is essential for situations where you do not have access to a computer or programming cable.

To begin manual programming, ensure your radio is turned on and in frequency mode rather than channel mode. Frequency mode allows you to input and adjust specific frequencies, while channel mode is used to access pre-programmed frequencies. Switching between these modes typically involves pressing the "VFO/MR" button on the radio.

Start by entering the desired frequency using the keypad. For example, if you want to program the frequency 146.520 MHz, simply key in 1-4-6-5-2-0. Ensure that the frequency you enter is within the operational range of your Baofeng radio and complies with local regulations.

Next, configure additional parameters such as the offset and tone. The offset, or repeater shift, is essential for communicating through repeaters. Repeaters are devices that receive a signal on one frequency and transmit it on another, allowing for extended communication range. To set the offset, press the "MENU" button, navigate to the "OFFSET" setting, and enter the appropriate value. Common offsets include 0.600 MHz for VHF and 5.000 MHz for UHF.

Tones are used for accessing repeaters that require a specific tone for activation, known as CTCSS (Continuous Tone-Coded Squelch System) or DCS (Digital-Coded Squelch). To set a tone, press the "MENU" button, find the

"T-CTCS" or "DCS" setting, and enter the required tone frequency. These tones ensure that your signal activates the repeater without interference from other transmissions on the same frequency.

Once all parameters are set, save the frequency to a memory channel for easy access. Press the "MENU" button, navigate to the "MEM-CH" setting, and select an available channel number. Confirm your selection by pressing the "MENU" button again, and the frequency, along with its parameters, will be saved to the chosen channel. You can now switch to channel mode and access your programmed frequency by selecting the corresponding channel number.

Repeat this process for each frequency you wish to program. Manual programming may take time, but it provides a hands-on understanding of your radio's functions and ensures you can program frequencies in the field without additional equipment.

Using Software for Programming

While manual programming is an essential skill, using software to program your Baofeng radio can save time and reduce errors, especially when dealing with multiple frequencies and complex settings. CHIRP is a popular, free, open-source software that simplifies the programming process and offers a user-friendly interface.

To begin using CHIRP, download and install the software from the official CHIRP website. Ensure you have a compatible programming cable that connects your Baofeng radio to your computer. These cables typically use a USB connection and are readily available from various retailers.

Once CHIRP is installed, connect your radio to your computer using the programming cable and turn on the radio. Open CHIRP and select the appropriate COM port that corresponds to your programming cable. This can usually be found under the "Port" dropdown menu in the "Radio" tab. Next, choose your radio model from the list provided.

Start by downloading the current configuration from your radio. This step ensures you have a backup of your existing settings and provides a foundation for further programming. In CHIRP, select "Download From Radio" from the "Radio" menu, follow the prompts, and save the downloaded data.

CHIRP allows you to easily add, modify, and organize frequencies. You can manually enter frequencies, tones, and other settings into the CHIRP interface or import data from various sources, such as frequency databases and CSV files. The software also enables you to configure additional features like power levels, step sizes, and scan lists, which might be cumbersome to set manually.

To add a new frequency, simply enter the details into the relevant fields in CHIRP's spreadsheet-like interface. Include information such as frequency, name, tone mode, tone frequency, offset, and power level. Repeat this process for each frequency you wish to program. You can also copy and paste data from spreadsheets or text files, streamlining the process further.

Once you have configured all desired frequencies and settings, upload the new configuration to your radio. In CHIRP, select "Upload To Radio" from the "Radio" menu and follow the prompts. Ensure your radio remains connected and powered on throughout the upload process. After the upload is complete, your radio will have the updated settings and programmed frequencies.

Using CHIRP or similar software not only speeds up the programming process but also provides a convenient way to manage and back up your radio's configuration. It is especially useful for maintaining multiple radios or frequently updating your settings.

Storing Frequencies

Storing frequencies in your Baofeng radio involves organizing them into channels and memory banks, allowing for quick access and efficient use. Understanding how to effectively store and manage these frequencies is crucial for streamlined communication.

Memory channels are individual storage slots within your radio where specific frequencies and their associated parameters are saved. These channels can be accessed quickly by switching to channel mode and selecting the desired channel number. Organizing frequencies into memory channels allows you to categorize them based on their use, such as emergency channels, local repeaters, or specific communication groups.

To store a frequency in a memory channel, follow the manual programming steps to enter the frequency and set the necessary parameters. Once everything is configured, save the frequency to an available memory channel. Each Baofeng model may have a different number of available memory channels, typically ranging from 128 to 200 channels.

It is important to assign meaningful names or labels to your memory channels. While Baofeng radios with basic displays may not support alphanumeric labels, more advanced models or using software like CHIRP can help you name channels for easy identification. Naming channels based on their purpose, location, or frequency band can help you quickly locate and access the desired channel during operation.

Memory banks are groups of memory channels organized into categories or lists. This feature is useful for segmenting frequencies based on different criteria, such as geographical regions, communication networks, or specific activities. By grouping related channels into memory banks, you can streamline your radio operation and ensure that you are always tuned to the appropriate frequencies for your current needs.

To set up memory banks, use CHIRP or similar programming software. In CHIRP, you can assign channels to specific banks by selecting the bank column and entering the bank number. You can then navigate through these banks on your radio, accessing only the channels within the selected bank. This organizational approach simplifies the management of numerous frequencies and ensures you are always prepared for various communication scenarios.

Priority channels are another useful feature for efficient communication. These channels are designated as high-priority and are scanned more frequently than other channels during scanning operations. This ensures that important frequencies are monitored more closely, allowing you to quickly respond to critical communications.

To set a priority channel, consult your radio's manual for specific instructions, as the process may vary between models. Typically, you will need to navigate to the desired channel and select it as a priority channel through the menu settings. Some radios allow for multiple priority channels, providing flexibility in monitoring essential frequencies.

By effectively storing and organizing frequencies into memory channels, memory banks, and priority channels, you can optimize your Baofeng radio's performance and ensure efficient communication. This organizational approach simplifies radio operation, allowing you to quickly access the necessary frequencies and stay prepared for any communication needs.

CHAPTER 6: Emergency Frequencies and Networks

Emergency Channels

During crisis, having immediate access to emergency frequencies is crucial. These channels are specifically designated for urgent communication and coordination during emergencies, such as natural disasters, accidents, or other critical situations. Knowing the national and local emergency frequencies, and how to access them with your Baofeng radio, can make a significant difference in your ability to stay informed and get help when needed.

National emergency frequencies are standardized across a country to ensure that everyone can communicate on the same channels during an emergency. In the United States, for example, the National Oceanic and Atmospheric Administration (NOAA) Weather Radio broadcasts continuous weather information on seven frequencies in the VHF band, ranging from 162.400 MHz to 162.550 MHz. These frequencies provide updates on severe weather conditions, natural disasters, and other critical information. Another important national frequency is the Citizen's Band (CB) radio emergency channel, typically channel 9 (27.065 MHz AM). While CB radios operate on different frequencies than Baofeng radios, it is still valuable to be aware of this channel for monitoring purposes if you have access to a CB radio. Channel 9 is widely recognized for emergency communication and is monitored by some local authorities and emergency services.

Local emergency frequencies vary based on the specific needs and regulations of each community. These frequencies are often used by public safety agencies, such as police, fire departments, and medical services. To find local emergency frequencies, consult local government websites, public safety agencies, or online frequency databases. Programming these frequencies into your Baofeng radio allows you to stay updated on local incidents and coordinate with nearby responders. In addition to dedicated emergency channels, amateur radio (HAM) operators often play a critical role in emergency communication. During disasters, HAM radio operators provide essential communication support, particularly when conventional communication infrastructure is compromised. Familiarize yourself with the emergency protocols of local HAM radio groups and consider obtaining an amateur radio license to actively participate in these networks.

Joining Local Radio Networks

Local radio networks, particularly those organized by amateur radio operators, are invaluable resources during emergencies. These networks, also known as nets, provide structured communication channels for sharing information, coordinating responses, and offering assistance. Joining local HAM radio networks enhances your ability to communicate effectively and contribute to community resilience during crises.

To find and join local radio networks, start by researching local amateur radio clubs and organizations. These groups often host regular nets on specific frequencies, providing a platform for members to practice communication skills and share information. Many clubs maintain websites or social media pages where they post net schedules and frequencies. Attending club meetings and participating in their activities is an excellent way to build connections and learn about local nets.

Once you have identified the frequencies and schedules of local nets, program them into your Baofeng radio. Typically, nets are conducted on VHF or UHF frequencies, depending on the range and coverage area. Some nets may operate on simplex frequencies (direct communication between radios), while others use repeaters to extend the range.

Participating in a net involves following specific protocols to ensure orderly and efficient communication. Nets are usually moderated by a Net Control Station (NCS), which manages the flow of communication and ensures that all participants have the opportunity to speak. When joining a net, wait for an appropriate break in the conversation, then identify yourself using your call sign and indicate your intention to join. For example, you might say, "Net Control, this is [Your Call Sign], requesting to join the net." During the net, follow the NCS's instructions and adhere to established protocols. Speak clearly and concisely, providing relevant information or updates as needed. Nets often have specific formats, such as check-ins, announcements, and open discussions. Familiarize yourself with these formats to ensure smooth participation. Many amateur radio groups organize emergency nets during crises. These emergency nets prioritize urgent communication and coordination, providing a critical link for information sharing and resource deployment. Participating in emergency nets requires a good understanding of emergency communication protocols and the ability to remain calm and focused under pressure.

Monitoring Emergency Services

Monitoring emergency services channels is a vital aspect of staying informed and prepared during crises. Public safety agencies, such as police, fire departments, and medical services, use specific frequencies for their communication needs. By scanning these channels, you can gain real-time insights into ongoing incidents, response efforts, and potential threats.

Most Baofeng radios come equipped with a scanning function, allowing you to search through a range of frequencies and detect active transmissions. To set up scanning, access the menu settings on your radio and configure the scan parameters, such as the frequency range and step size. Initiate the scan and listen for transmissions from emergency services. When scanning for police channels, you will primarily be looking for frequencies in the VHF and UHF bands. Police departments often use trunked radio systems, which involve multiple frequencies and digital modulation techniques. While Baofeng radios are not capable of directly decoding trunked systems, they can still pick up conventional analog frequencies used by some police departments. Resources like online frequency databases and scanner enthusiast communities can help identify the specific frequencies used by your local police.

Fire departments also operate on VHF and UHF frequencies, often using repeaters to extend their communication range. These channels are essential for monitoring incidents such as fires, hazardous material spills, and rescue operations. Fire department communication tends to be more straightforward than police communication, making it easier to follow and understand the nature of the incidents being addressed.

Medical services, including ambulance and emergency medical technicians (EMTs), use specific frequencies for dispatch and coordination. These channels provide critical information on medical emergencies, patient transport, and hospital coordination. Monitoring these frequencies can help you stay aware of medical incidents in your area and prepare for potential impacts.

National and regional emergency communication networks also play a crucial role. For example, the National Weather Service (NWS) operates NOAA Weather Radio, providing continuous updates on weather conditions and warnings. Monitoring these broadcasts on your Baofeng radio keeps you informed about severe weather events, such as hurricanes, tornadoes, and floods, allowing you to take timely protective measures.

During major disasters, agencies like the Federal Emergency Management Agency (FEMA) and the Red Cross may establish temporary communication networks to coordinate relief efforts. These networks often use designated frequencies and protocols to ensure efficient communication among responders. Staying informed about these frequencies and monitoring their transmissions can provide valuable insights into the broader response efforts and available resources.

While monitoring emergency services channels is beneficial, it is important to use this capability responsibly. Avoid interfering with official communication and follow all legal regulations regarding frequency use and monitoring. Being a well-informed and responsible radio operator enhances your ability to contribute to community resilience and support emergency response efforts.

By understanding and utilizing emergency frequencies and networks, you can significantly enhance your preparedness and response capabilities during crises. Knowing how to access national and local emergency channels, joining local radio networks, and effectively monitoring emergency services provides a comprehensive approach to staying informed and connected. This knowledge and capability ensure you are well-equipped to handle emergencies, contribute to community resilience, and support coordinated response efforts.

CHAPTER 7: Advanced Communication Techniques

Repeater Use

Repeater use is a fundamental aspect of advanced communication techniques, particularly for extending the range and coverage of your Baofeng radio. Repeaters are devices that receive signals on one frequency and retransmit them on another frequency, typically at higher power and from an elevated location. By accessing repeaters strategically positioned throughout your area, you can significantly enhance your communication range and clarity.

To use a repeater with your Baofeng radio, you need to know the repeater's frequency, offset, and tone (if required). Many repeaters operate on standard frequencies allocated for amateur radio use, such as the 2-meter (144-148 MHz) and 70-centimeter (420-450 MHz) bands. These frequencies are organized into repeater pairs, with a specific offset between the input (receive) and output (transmit) frequencies.

When programming a repeater into your Baofeng radio, ensure you set the correct offset and tone settings. The offset is typically provided by the repeater operator or listed in frequency directories. Common offsets for VHF repeaters are +0.600 MHz or -0.600 MHz, while UHF repeaters often use +5.000 MHz or -5.000 MHz. Additionally, some repeaters require a specific tone (CTCSS or DCS) for access, which must be configured in your radio's settings.

To access a repeater, tune your Baofeng radio to the repeater's input frequency. Once on the input frequency, transmit your signal as you would normally. The repeater receives your signal, amplifies it, and retransmits it on its output frequency, extending your communication range significantly. When receiving signals from the repeater, your radio automatically switches to the repeater's output frequency, allowing you to hear the transmitted messages.

Using repeaters effectively requires an understanding of repeater coverage areas and antenna placement. Repeaters are typically located on elevated sites, such as mountain peaks or tall buildings, to maximize their coverage. Before using a repeater, check its coverage map or consult local amateur radio clubs for information on its range and reliability. Experiment with different antenna configurations and locations to optimize your signal strength and access to repeaters in your area.

Signal Boosting

Signal boosting techniques are essential for improving the range, clarity, and reliability of your Baofeng radio communication. Whether operating in challenging environments or attempting long-distance communication, signal boosting techniques help overcome obstacles and enhance overall performance.

One of the simplest ways to boost your signal is by using a high-quality antenna. The antenna is one of the most critical components of your radio system and significantly influences signal propagation. Upgrading to a longer, higher-gain antenna can increase your radio's effective range and improve its ability to transmit and receive signals in various conditions. Consider antennas specifically designed for the frequency bands used by your Baofeng radio, such as VHF/UHF dual-band antennas or directional antennas for focused coverage. Optimizing antenna placement is another effective signal boosting technique. Mounting your antenna in a high, unobstructed location, such as on a rooftop or mast, maximizes its line-of-sight coverage and minimizes signal attenuation from obstacles like buildings, trees, and terrain. Experiment with different mounting heights and orientations to find the optimal setup for your specific communication needs.

Using a signal amplifier, also known as a power amplifier or booster, can further enhance your radio's transmission capabilities. Signal amplifiers increase the power output of your radio, allowing for stronger signals and extended communication range. However, it is essential to use amplifiers responsibly and within legal limits to avoid interference with other users and regulatory violations. Check local regulations and licensing requirements before using a signal amplifier with your Baofeng radio. In urban environments or areas with high levels of electromagnetic interference, noise reduction techniques can help improve signal clarity and readability. Noise-canceling headphones or external filters can suppress background noise and unwanted interference, allowing you to focus on incoming signals. Additionally, adjusting your radio's squelch settings can help eliminate weak or noisy signals, improving overall communication quality.

Encryption and Privacy

Encryption and privacy features are critical for securing your communications and protecting sensitive information from unauthorized interception or eavesdropping. While Baofeng radios typically do not have built-in encryption capabilities, there are alternative methods for enhancing privacy and security. One approach is to use digital voice modes, such as Digital Mobile Radio (DMR) or Digital Private Mobile Radio (dPMR), which offer inherent encryption features. These digital modes use advanced modulation techniques and encryption algorithms to encode voice transmissions, making them resistant to unauthorized monitoring. However, using digital modes requires compatible radios and coordination with other users on the same network.

Another option is to use encryption devices or modules specifically designed for analog radios. These encryption devices connect between your Baofeng radio and microphone, encrypting outgoing transmissions and decrypting incoming ones. While effective at providing privacy, encryption devices can be costly and may require additional setup and configuration.

In situations where encryption is not feasible or practical, implementing privacy measures such as code words, language codes, or prearranged signals can help prevent unauthorized listeners from understanding your communications. Establishing clear communication protocols and using discretion when discussing sensitive information can also minimize the risk of interception.

Regardless of the method used, it is essential to understand the legal and regulatory implications of encrypting radio communications. In many countries, encryption of certain radio frequencies or services is restricted or prohibited, and unauthorized use of encryption may result in fines or penalties. Consult local regulations and seek guidance from legal experts or regulatory authorities to ensure compliance with applicable laws.

Maintaining privacy and security in radio communication requires a combination of technical solutions, operational practices, and regulatory compliance. By employing encryption and privacy measures, you can safeguard your communications and protect sensitive information from unauthorized access, ensuring the integrity and confidentiality of your transmissions.

CHAPTER 8: Guerrilla Communication Strategies

Introduction to Guerrilla Tactics

Guerrilla communication strategies involve the use of unconventional and resourceful methods to establish and maintain communication channels, often in challenging or hostile environments. Derived from guerrilla warfare tactics, which emphasize flexibility, adaptability, and asymmetrical approaches, guerrilla communication tactics prioritize agility, secrecy, and resilience. These strategies are particularly relevant in situations where traditional communication infrastructure is unavailable, compromised, or monitored, such as during emergencies, conflicts, or oppressive regimes.

The principles of guerrilla communication tactics are rooted in the history of irregular warfare and resistance movements. Throughout history, guerrilla fighters and resistance groups have relied on clandestine communication methods to coordinate operations, disseminate information, and maintain morale. From secret codes and hidden messages to covert radio transmissions and clandestine networks, guerrilla communication has played a vital role in asymmetric conflicts and resistance movements worldwide.

Guerrilla communication strategies encompass a diverse range of techniques and technologies, leveraging advances in telecommunications, cryptography, and improvisation. These strategies are characterized by their adaptability to changing circumstances, reliance on stealth and discretion, and emphasis on decentralized and resilient networks. By understanding and employing guerrilla communication tactics, individuals and groups can enhance their ability to communicate securely and effectively in challenging environments.

Stealth Communication Techniques

Stealth communication techniques are essential components of guerrilla communication strategies, enabling users to transmit and receive information discreetly and avoid detection by adversaries or monitoring authorities. Whether operating in hostile environments, conducting covert operations, or simply seeking to maintain privacy, mastering stealth communication techniques is critical for safeguarding sensitive information and ensuring operational security.

One stealth communication technique is the use of low-power transmissions and narrowband modes to minimize the risk of detection. By reducing the power output of their radios and utilizing narrowband modulation schemes, users can limit the range and visibility of their transmissions, making them less susceptible to interception or direction finding. Additionally, employing frequency-hopping spread spectrum (FHSS) or frequency-shift keying (FSK) modulation techniques can further obscure transmissions and make them harder to intercept or jam. Another stealth communication technique is the use of covert or disguised antennas to conceal radio installations and minimize their visibility. Covert antennas, such as wire antennas hidden in trees or buildings, or magnetic loop antennas concealed within structures or vehicles, can help mitigate the risk of detection by blending into the surrounding environment. Additionally, deploying directional antennas with narrow beamwidths can focus transmissions and reduce the likelihood of interception by unauthorized listeners.

In urban environments or areas with high levels of electronic surveillance, using signal masking or camouflage techniques can help conceal radio transmissions and avoid detection. Signal masking involves transmitting in proximity to other electromagnetic sources, such as power lines, electrical equipment, or electronic devices, to mask the radio signal and make it harder to isolate or locate. Camouflage techniques, such as using radio frequency (RF) shielding materials or enclosures, can also help conceal radio installations and minimize their detectability. Operational security (OPSEC) is another critical aspect of stealth communication, focusing on the protection of sensitive information and the prevention of adversary exploitation. Practicing good OPSEC involves minimizing the disclosure of sensitive information, limiting communication to essential messages, and avoiding patterns or routines that could be exploited by adversaries. By maintaining a low profile, exercising caution when discussing sensitive topics, and adhering to strict operational security protocols, users can mitigate the risk of compromise and maintain the secrecy of their communications.

Improvising Antennas and Power Sources

The ability to improvise antennas and power sources from available materials is essential for establishing communication capabilities in remote or austere environments. Whether operating in the wilderness, urban areas, or during emergencies, improvising antennas and power sources allows users to overcome logistical challenges and maintain communication resilience.

Field-expedient antennas can be constructed from a variety of materials, including wire, coaxial cable, metal rods, and even natural elements like trees or foliage. Improvised antennas, such as long wire antennas, dipole antennas, or ground plane antennas, can be fashioned quickly and deployed in temporary or semi-permanent configurations to establish communication links. By understanding basic antenna principles and experimenting with different designs and materials, users can optimize antenna performance and adapt to changing operational requirements.
Power sources for portable radios can be improvised from a range of readily available materials, including batteries, solar panels, hand-crank generators, and vehicle power sources. In remote or off-grid environments, solar panels offer a renewable and sustainable

power solution, harnessing solar energy to recharge batteries or power radios directly. Hand-crank generators provide a portable and self-sufficient power source, allowing users to generate electricity manually when other options are unavailable.

In urban environments or during emergencies, scavenging for power sources such as vehicle batteries, mains power outlets, or alternative energy sources can provide temporary power for radios and communication equipment. Vehicle batteries, in particular, offer a reliable and accessible power source, providing ample voltage and capacity for extended communication operations. By improvising power sources from available materials and resources, users can maintain communication capabilities in challenging circumstances and ensure operational continuity.

By mastering stealth communication techniques and improvising antennas and power sources, individuals and groups can enhance their ability to communicate securely and effectively in diverse environments. Whether operating in hostile environments, conducting covert operations, or simply seeking to maintain privacy, guerrilla communication strategies provide a versatile and adaptable framework for establishing resilient communication networks and safeguarding sensitive information.

CHAPTER 9: Setting Up a Base Station

Choosing a Location

Setting up a base station requires careful consideration of the location to ensure optimal performance and reliability. The ideal location for a base station is one that provides clear line-of-sight communication, minimal interference, and adequate protection from environmental elements.

When selecting a location for your base station, consider the following factors:

Elevation: Choose a location with sufficient elevation to maximize line-of-sight communication and minimize obstructions. Elevated sites, such as hills, rooftops, or towers, offer improved signal propagation and coverage compared to low-lying areas.

Accessibility: Ensure the chosen location is easily accessible for installation, maintenance, and operation. Accessibility is particularly important for deploying equipment, running power and communication cables, and performing routine checks and adjustments.

Interference: Avoid locations with significant electromagnetic interference (EMI) sources, such as power lines, electronic equipment, or radio frequency (RF) transmitters. Interference can degrade signal quality, increase noise levels, and compromise communication reliability.

Security: Consider the security of the chosen location to protect equipment from theft, vandalism, or tampering. Secure sites, such as fenced compounds, locked rooms, or monitored facilities, offer added protection against unauthorized access and ensure the integrity of your base station setup.

Environmental Factors: Assess environmental factors such as exposure to extreme weather conditions, precipitation, and temperature fluctuations. Choose a location that provides shelter from harsh weather elements and minimizes the risk of equipment damage or degradation over time.

By carefully evaluating these factors and selecting a suitable location for your base station, you can maximize communication range, reliability, and performance while minimizing potential challenges and vulnerabilities.

Equipment and Setup

Once you have chosen a suitable location for your base station, it's time to gather the necessary equipment and set up your station for operation. The equipment required for a base station setup varies depending on your communication needs, frequency bands, and deployment scenario.
However, common components of a base station setup include:

Radio Transceiver: Select a high-quality radio transceiver suitable for your intended use and frequency bands. Choose a model with features such as adjustable output power, frequency agility, and compatibility with external antennas and accessories.

Antenna: Install an appropriate antenna for your base station setup, considering factors such as frequency bands, gain, polarization, and directional characteristics. Choose an antenna design optimized for your communication requirements, whether omnidirectional for wide coverage or directional for focused transmission.

Power Supply: Ensure reliable power for your base station by using a stable and sufficient power supply. Depending on your location and available resources, options for power supply include mains electricity, batteries, solar panels, or generators. Consider backup power solutions to maintain operation during power outages or emergencies.

Mounting Hardware: Securely mount your radio transceiver, antenna, and associated equipment using suitable mounting hardware.

Choose mounting solutions designed for durability, weather resistance, and compatibility with your installation location, such as roof mounts, pole mounts, or wall brackets.

Cables and Connectors: Use high-quality cables and connectors to connect your radio transceiver, antenna, and power supply components. Select cables with appropriate lengths, types, and impedance matching to minimize signal loss, interference, and degradation.

Grounding System: Implement a proper grounding system to protect your base station equipment from electrical surges, static discharge, and lightning strikes. Install grounding rods, conductors, and bonding connections according to industry standards and local regulations to ensure effective grounding and safety.

Accessories: Consider additional accessories to enhance the functionality and performance of your base station setup, such as lightning arrestors, surge protectors, antenna tuners, filters, and remote control systems. Choose accessories compatible with your equipment and deployment requirements to optimize system operation and reliability.

Once you have gathered the necessary equipment, follow these steps to set up your base station:
1. Preparation: Prepare the installation site by clearing obstructions, assessing environmental conditions, and ensuring accessibility for equipment deployment.

2. Installation: Install the radio transceiver, antenna, power supply, and associated equipment according to manufacturer guidelines, installation instructions, and best practices for your specific setup.

3. Configuration: Configure the radio transceiver settings, antenna parameters, power supply voltage, and other system parameters to optimize performance and ensure compatibility with your communication objectives.

4. Testing: Conduct thorough testing and validation of your base station setup to verify functionality, performance, and reliability. Test communication range, signal quality, and system stability under various conditions and scenarios.

5. Optimization: Fine-tune and optimize your base station setup based on testing results, feedback, and operational requirements. Adjust antenna positioning, power levels, frequency settings, and other parameters as needed to achieve optimal performance and coverage.

By following these steps and guidelines, you can set up a robust and reliable base station capable of providing effective communication capabilities for your intended applications.

Maintaining Your Base Station

Maintaining your base station is essential for ensuring continued performance, reliability, and longevity of your communication infrastructure. Regular maintenance and upkeep help prevent equipment failures, optimize system performance, and minimize downtime.

Here are some maintenance tasks and best practices for maintaining your base station:

Scheduled Inspections: Conduct routine inspections of your base station equipment to check for signs of wear, damage, or degradation. Inspect cables, connectors, antennas, mounting hardware, and grounding systems for corrosion, loose connections, or physical damage.

Cleaning and Decontamination: Clean and decontaminate your base station equipment regularly to remove dust, dirt, debris, and environmental contaminants. Use mild detergents, solvents, or cleaning solutions appropriate for electronic equipment and avoid abrasive materials that could damage surfaces or components.

Antenna Maintenance: Inspect and maintain your antenna regularly to ensure optimal performance and longevity. Check for signs of corrosion, damage, or misalignment, and clean antenna elements, radomes, and feedlines as needed to maintain signal integrity and efficiency.

Grounding System Inspection: Verify the integrity and effectiveness of your grounding system by inspecting grounding rods, conductors, bonding connections, and earth electrodes. Ensure that grounding components are properly installed, securely attached, and free from corrosion or damage.

Power Supply Monitoring: Monitor the performance and condition of your power supply components, including batteries, solar panels, and generators. Check battery voltage levels, solar panel output, and generator operation regularly to detect any issues or anomalies that could affect system reliability.

Software Updates and Maintenance: Keep your radio transceiver firmware, software, and configuration settings up to date by applying software updates, patches, and security fixes as recommended by the manufacturer. Regularly review and optimize radio settings, channel configurations, and system parameters to adapt to changing operational requirements.

Environmental Protection: Protect your base station equipment from environmental hazards, such as extreme weather conditions, temperature fluctuations, moisture, and pollutants. Install weatherproof enclosures, protective covers, or shelter structures to shield sensitive equipment from the elements and minimize exposure to environmental stressors.

Emergency Preparedness: Develop and implement contingency plans and emergency procedures to respond to equipment failures, power outages, or other unforeseen events. Maintain backup power supplies, spare parts, replacement components, and alternative communication options to ensure continuity of operations during emergencies.

CHAPTER 10: Mobile and Field Operations

In mobile and field operations, the ability to effectively utilize your radio equipment is paramount for maintaining communication capabilities while on the move or deployed in remote locations. Whether conducting field exercises, emergency response missions, or outdoor activities, understanding how to optimize your radio performance and adapt to changing environments is essential for successful operations.

Using Your Radio in the Field

Using your radio effectively in the field requires a combination of technical proficiency, situational awareness, and operational discipline. **Here are some tips for maximizing the effectiveness of your field communication:**

Frequency Selection: Choose appropriate frequencies for your communication needs, considering factors such as range, interference, and regulatory constraints. Use frequency bands and channels allocated for your intended use, such as amateur radio bands for non-commercial communication or public safety frequencies for emergency operations.

Channel Management: Organize and prioritize your radio channels based on their importance and relevance to your mission or activities. Assign dedicated channels for specific purposes, such as tactical communication, command coordination, or emergency response, and establish protocols for channel selection and usage.

Clear and Concise Communication: Practice clear and concise communication techniques to convey information effectively and efficiently. Use standardized procedures, terminology, and call signs to facilitate rapid and accurate communication, particularly in fast-paced or high-stress situations.

Situational Awareness: Maintain situational awareness of your surroundings, operational objectives, and potential hazards to anticipate communication requirements and respond effectively to changing conditions. Monitor radio traffic, listen for relevant information, and adapt your communication strategy as needed to support mission objectives.

Radio Etiquette: Follow established radio etiquette and protocols to ensure professionalism, courtesy, and orderliness in your communication exchanges. Avoid unnecessary transmissions, interruptions, or chatter that could disrupt operational activities or compromise security.

Antenna Orientation: Orient your radio antenna for optimal signal propagation and reception, taking into account factors such as terrain, obstacles, and polarization. Position the antenna vertically for maximum range and coverage in open environments, or adjust its orientation to minimize multipath interference and signal degradation.

Battery Management: Manage your radio battery effectively to maintain sufficient power for extended operations. Monitor battery levels regularly, conserve power by minimizing unnecessary transmissions or features, and carry spare batteries or backup power sources to ensure continuity of communication.

By implementing these strategies and techniques, you can enhance the effectiveness and reliability of your field communication operations, enabling seamless coordination and collaboration in dynamic and challenging environments.

Vehicle Installation

Installing your radio equipment in a vehicle provides mobile communication capabilities for on-the-go operations, transportation logistics, and field deployments. Whether traveling to remote locations, conducting mobile patrols, or supporting emergency response missions, setting up your radio in a vehicle enhances situational awareness, coordination, and response capabilities.

Here are some key considerations for vehicle installation:

Mounting Location: Choose a suitable mounting location for your radio equipment within the vehicle, considering factors such as accessibility, visibility, and ergonomics. Common mounting locations include the dashboard, center console, or overhead console, depending on available space and vehicle layout.

Mounting Hardware: Use secure and durable mounting hardware to affix your radio equipment to the vehicle, ensuring stability, vibration resistance, and shock absorption. Select mounting solutions designed for automotive applications, such as vehicle-specific mounting brackets, console mounts, or dash mounts, to minimize installation complexity and ensure compatibility.

Power Connection: Connect your radio equipment to the vehicle's electrical system to provide power for operation. Use appropriate wiring, fuses, and connectors to establish a reliable electrical connection, and ensure compatibility with the vehicle's voltage, current, and power distribution system.

Antenna Placement: Install external antennas on the vehicle to maximize communication range and coverage. Choose antenna mounts and locations that offer clear line-of-sight visibility, minimal interference, and optimal signal propagation. Position antennas vertically for maximum performance and avoid mounting them near metallic or obstructive objects that could affect radiation patterns or signal quality.

Coaxial Cable Routing: Route coaxial cables from the radio equipment to the antennas through the vehicle's interior or exterior channels, ensuring proper routing, strain relief, and weather sealing. Secure cables to prevent chafing, rubbing, or damage from sharp edges or moving parts, and use cable management accessories such as grommets, clips, or conduits to organize and protect the wiring.

Grounding: Ground the radio equipment and antennas to the vehicle chassis to ensure electrical safety, RF performance, and system integrity. Establish a solid ground connection using grounding straps, conductive bonding, or dedicated grounding points, and verify continuity and resistance to minimize ground loop interference and electrical hazards.

By following these guidelines and best practices, you can install your radio equipment in a vehicle securely, efficiently, and safely, enabling mobile communication capabilities for a wide range of operational scenarios and mission requirements.

Field Antennas and Power Sources

In field operations, portable antennas and power sources are essential for establishing communication capabilities in remote or austere environments where traditional infrastructure is unavailable or impractical. Whether operating in the wilderness, conducting outdoor activities, or supporting emergency response missions, portable solutions for antennas and power sources enable extended communication range, flexibility, and resilience. Here are some options for field antennas and power sources:

Portable Antennas:

- **Wire Antennas:** Construct wire antennas from lightweight, flexible wire or cordage to deploy in temporary or semi-permanent configurations. Wire antennas, such as dipole antennas, long wire antennas, or end-fed antennas, are versatile and easy to deploy, requiring minimal space and equipment.

- **Compact Antennas:** Use compact antennas designed for portable or mobile applications, such as whip antennas, telescoping antennas, or collapsible antennas. Compact antennas offer convenience, portability, and quick deployment, making them ideal for field operations and outdoor activities.

- **Directional Antennas:** Deploy directional antennas, such as Yagi antennas, log-periodic antennas, or panel antennas, to focus transmission and reception in specific directions for increased range and coverage. Directional antennas are particularly useful for point-to-point communication, long-distance links, and signal directionality control.

Power Sources:

- **Battery Packs:** Use rechargeable battery packs or portable power banks to supply electrical power for radios, accessories, and communication equipment in the field. Battery packs offer mobility, versatility, and convenience, allowing for extended operation without reliance on external power sources.

- **Solar Panels:** Harness solar energy with portable solar panels to recharge batteries, power radios, or operate communication equipment in off-grid environments. Solar panels provide renewable and sustainable power solutions for field operations, outdoor activities, and emergency response missions, reducing reliance on fossil fuels and grid electricity.

- **Hand-Crank Generators:** Generate electricity manually with hand-crank generators or dynamo chargers to power radios, charge batteries, or operate communication devices in remote or emergency situations. Hand-crank generators offer a self-sufficient power source, requiring no external fuel or infrastructure for operation.

- **Vehicle Power Sources:** Utilize vehicle power sources, such as vehicle batteries, alternators, or cigarette lighter sockets, to supply electrical power for radios and communication equipment while on the move. Vehicle power sources provide reliable and accessible power options for mobile operations, transportation logistics, and field deployments.

By leveraging portable antennas and power sources, individuals and teams can establish communication capabilities in challenging environments, maintain operational connectivity, and support mission-critical tasks effectively. Whether operating in the field, conducting outdoor activities, or responding to emergencies, portable solutions for antennas and power sources enable mobility, flexibility

CHAPTER 11: Survival Communication Scenarios

The ability to maintain communication channels can be a matter of life and death. Whether facing natural disasters, man-made emergencies, or extreme societal collapse, having reliable means of communication is essential for coordinating rescue efforts, accessing vital information, and ensuring personal safety. This chapter explores various survival communication scenarios, including natural disasters, man-made emergencies, and SHTF (Sht Hits The Fan) scenarios, and discusses strategies for effective communication in each situation.

Natural Disasters

Natural disasters pose significant challenges to communication infrastructure, often causing widespread damage to traditional communication networks and infrastructure. From hurricanes and earthquakes to floods and wildfires, natural disasters can disrupt power supply, damage communication towers, and server communication links, leaving affected areas isolated and vulnerable. In such scenarios, alternative communication methods become essential for coordinating emergency response efforts, disseminating critical information, and providing assistance to affected populations.

Hurricanes

Hurricanes are powerful tropical storms characterized by strong winds, heavy rain, and storm surges, posing significant risks to coastal communities and inland areas. During hurricanes, communication networks may be compromised due to power outages, flooding, and infrastructure damage. Alternative communication methods, such as satellite phones, amateur radio networks, and mesh networks, can provide lifelines for emergency responders and affected communities, enabling coordination, information exchange, and resource allocation.

Earthquakes

Earthquakes can cause widespread destruction and disruption to communication infrastructure, particularly in densely populated urban areas. Collapsed buildings, damaged roads, and disrupted power supply can impede communication and rescue operations, hampering efforts to locate survivors and deliver assistance. Portable communication devices, such as handheld radios, satellite messengers, and personal locator beacons (PLBs), can facilitate communication and coordination among rescuers and affected individuals, improving response efficiency and effectiveness.

Floods

Flooding can inundate low-lying areas, submerge infrastructure, and disrupt communication networks, making it challenging to assess damage, coordinate evacuations, and deliver emergency services. In flood-affected regions, communication tools such as waterproof radios, marine VHF radios, and buoyant communication devices are essential for maintaining contact with rescue teams, coordinating evacuations, and monitoring flood conditions in real-time.

Wildfires
Wildfires can spread rapidly, engulfing vast areas of vegetation and threatening homes, communities, and natural habitats. In wildfire-prone areas, communication systems may be compromised by smoke, ash, and heat, limiting visibility and signal transmission. Wireless mesh networks, drone-based communication relays, and satellite-based imaging systems can provide situational awareness, communication support, and firefighting coordination in wildfire-affected regions, helping to contain fires and protect lives and property.

By leveraging alternative communication methods and technologies, emergency responders, disaster relief organizations, and affected communities can overcome communication challenges posed by natural disasters, enhance resilience, and mitigate the impact of adverse events on public safety and well-being.

Man-Made Disasters

Man-made disasters, including terrorist attacks, industrial accidents, and civil unrest, can disrupt communication networks, disrupt social order, and pose significant threats to public safety and security. Whether perpetrated by individuals, groups, or state actors, man-made disasters require effective communication strategies and response measures to mitigate the impact and protect lives and property.

Terrorist Attacks:
Terrorist attacks, such as bombings, shootings, and cyber-attacks, can target critical infrastructure, public spaces, and civilian populations, causing mass casualties and widespread panic. During terrorist incidents, communication networks may be overwhelmed by emergency calls, misinformation, and rumors, hindering coordination and response efforts.

Secure communication channels, encrypted messaging platforms, and crisis communication protocols can help authorities communicate securely, disseminate accurate information, and coordinate emergency response activities, minimizing the impact of terrorist attacks and restoring public confidence.

Civil Unrest

Civil unrest, protests, and demonstrations can escalate into violent confrontations, property damage, and social disorder, posing challenges to law enforcement, emergency services, and public safety agencies. Communication blackout, internet shutdowns, and social media censorship may be used by authorities to control information flow and suppress dissent, exacerbating tensions and limiting transparency. Decentralized communication networks, peer-to-peer messaging apps, and mesh networking technologies can provide alternative channels for communication and coordination among protesters, activists, and concerned citizens, enabling them to organize, mobilize, and advocate for social change in challenging environments.

Industrial Accidents

Industrial accidents, such as chemical spills, explosions, and hazardous material releases, can pose significant risks to public health, environmental safety, and community well-being. In the event of industrial accidents, communication systems may be overwhelmed by emergency calls, evacuation orders, and hazardous material warnings, impeding response efforts and exacerbating confusion. Two-way radios, inter-agency communication systems, and incident command protocols can facilitate coordination and collaboration among emergency responders, industrial facilities, and regulatory agencies, enabling timely and effective response to industrial accidents and minimizing the impact on human health and the environment.

By implementing robust communication plans, training personnel, and leveraging technology solutions, stakeholders can enhance preparedness, response, and recovery capabilities in the face of man-made disasters, safeguarding public safety, and ensuring resilience in the face of adversity.

SHTF Scenarios

SHTF (Sht Hits The Fan) scenarios refer to extreme situations characterized by total societal collapse, breakdown of law and order, and widespread chaos and disorder. While such scenarios are rare and unlikely, they require careful consideration of communication strategies and preparedness measures to ensure survival and security in extreme circumstances.

Communication Blackout

In SHTF scenarios, communication blackout may occur due to infrastructure damage, electromagnetic pulse (EMP) effects, or deliberate interference by hostile actors. In the absence of traditional communication networks, alternative communication methods such as handheld radios, signal mirrors, and Morse code can be used to maintain contact with family members, neighbors, and trusted allies, facilitating coordination, information exchange, and mutual assistance.

Community Networks

Building resilient community networks and mutual aid groups is essential for collective survival and security in SHTF scenarios. By establishing communication protocols, sharing resources, and coordinating activities, communities can strengthen solidarity, foster cooperation, and mitigate the impact of societal collapse on individual and collective well-being. Community-based radio networks, citizen band (CB) radios, and neighborhood watch programs can serve as vital lifelines for communication and collaboration among residents, enabling them to address challenges, solve problems, and support each other in times of crisis.

Survivalist Retreats

Survivalist retreats, also known as bug-out locations or safe havens, are fortified shelters or remote hideaways designed to provide refuge and protection in SHTF scenarios. Equipped with communication equipment, emergency supplies, and defensive measures, survivalist retreats enable individuals and families to weather the storm, defend against threats, and maintain communication with the outside world. Shortwave radios, satellite phones, and encrypted messaging apps can be used to establish communication links with trusted contacts, gather intelligence, and coordinate response actions, ensuring the security and survival of retreat occupants.

HAM Radio Networks

Amateur radio, or HAM radio, networks play a crucial role in SHTF scenarios, providing long-distance communication capabilities independent of traditional infrastructure. HAM radio operators, or "hams," can establish communication links across vast distances using high-frequency (HF) bands, repeaters, and satellite relays, enabling information exchange, emergency assistance, and community support in times of crisis. By obtaining a HAM radio license, learning Morse code, and participating in radio drills and exercises, individuals can acquire essential skills and resources for effective communication in SHTF scenarios, enhancing their resilience and self-reliance in the face of adversity.

Clandestine Communication

In SHTF scenarios where privacy and security are paramount, clandestine communication methods may be employed to avoid detection and interception by hostile actors or oppressive regimes. Techniques such as steganography, encryption, and dead drops can be used to conceal messages, protect sensitive information, and maintain operational security in hostile environments. By adopting operational security (OPSEC) measures, practicing discretion, and minimizing electronic footprint, individuals and groups can reduce the risk of surveillance, infiltration, and compromise, preserving their autonomy and anonymity in clandestine communication activities.

Information Warfare
In SHTF scenarios characterized by information warfare, propaganda, and psychological operations, critical thinking, media literacy, and discernment are essential for navigating the information landscape and distinguishing fact from fiction. By critically evaluating sources, cross-referencing information, and verifying claims, individuals can avoid manipulation, propaganda, and disinformation campaigns, empowering them to make informed decisions and take effective action in challenging circumstances. Skepticism, skepticism, and skepticism should be the watchwords of survival in the information age, enabling individuals to question, challenge, and verify information before accepting it as truth.

Survival communication scenarios present unique challenges and opportunities for individuals, communities, and organizations to adapt, innovate, and collaborate in the face of adversity. By understanding the dynamics of natural disasters, man-made emergencies, and extreme societal collapse, and by implementing effective communication strategies and preparedness measures, stakeholders can enhance their resilience, safeguard their security, and ensure their survival in even the most challenging circumstances. Through proactive planning, training, and cooperation, we can build a more resilient and prepared society capable of weathering the storm and emerging stronger on the other side.

CHAPTER 12: Prepping Essentials

Having a comprehensive plan and the necessary supplies can make all the difference in ensuring the safety and well-being of yourself and your loved ones. This chapter delves into the essentials of prepping, covering topics such as building a comprehensive prepper's kit, creating a family communication plan, and implementing long-term survival strategies to sustain communication over prolonged periods of uncertainty.

Building a Comprehensive Prepper's Kit

A well-equipped prepper's kit is the cornerstone of emergency preparedness, providing the essential gear and supplies needed to weather various crises and emergencies. When assembling your kit, consider the following categories of items:

Food and Water
Stockpile non-perishable food items such as canned goods, dry goods, and freeze-dried meals

to sustain yourself and your family during emergencies. Additionally, store an ample supply of clean drinking water, either in bottled form or through water purification methods such as filtration, boiling, or chemical treatment.

Shelter and Clothing

Include items to provide shelter and protect against the elements, such as tents, tarps, sleeping bags, and warm clothing. Consider the climate and terrain of your area when selecting shelter and clothing options to ensure suitability for your needs.

First Aid and Medical Supplies

Pack a comprehensive first aid kit containing bandages, wound dressings, antiseptics, medications, and medical tools to address injuries and illnesses during emergencies. Additionally, consider any specific medical needs or prescriptions required by members of your household.

Tools and Equipment

Include essential tools and equipment for survival and self-sufficiency, such as a multi-tool, flashlight, batteries, fire-starting materials, and a portable stove or cooking device. These items can be invaluable for performing tasks, making repairs, and maintaining comfort and safety in emergency situations.

Communication Devices
Incorporate communication devices into your prepper's kit to maintain contact with emergency services, family members, and other individuals during crises. Portable radios, satellite phones, and two-way communication devices can enable communication when traditional methods are unavailable or unreliable.

Personal Protection and Security
Prioritize personal protection and security by including items such as self-defense tools, protective gear, and emergency signaling devices in your prepper's kit. Consider the potential

threats and risks in your environment and prepare accordingly to mitigate them effectively.

Documentation and Identification
Keep important documents, identification, and contact information in waterproof and portable containers as part of your prepper's kit. Include copies of vital records, insurance policies, passports, and emergency contacts to facilitate communication and assistance during emergencies.

By assembling a comprehensive prepper's kit tailored to your specific needs and circumstances, you can enhance your readiness and resilience in the face of unforeseen emergencies and challenges.

Creating a Family Communication Plan

A family communication plan is essential for keeping loved ones informed, connected, and safe during emergencies, disasters, or crises. By establishing clear communication protocols, designated meeting points, and emergency contacts, you can ensure that everyone in your household knows what to do and where to go in case of an emergency. When creating a family communication plan, consider the following steps:

Identify Emergency Contacts: Compile a list of emergency contacts, including family members, friends, neighbors, and emergency services, and distribute copies to each household member. Ensure that everyone knows how to reach these contacts using multiple communication methods, such as phone calls, text messages, or social media.

Designate Meeting Points: Establish designated meeting points both inside and outside your home, as well as in your neighborhood or community, where family members can gather in case of evacuation or separation. Choose easily identifiable landmarks or locations that are accessible and safe for everyone to reach.

Establish Communication Protocols: Define communication protocols for different types of emergencies, specifying which communication methods to use, when to check in, and what information to share. Ensure that everyone knows how to use communication devices and apps effectively and understands the importance of staying connected during crises.

Practice Emergency Drills: Conduct regular emergency drills and exercises with your family to practice communication protocols, evacuation procedures, and emergency responses. Simulate various scenarios, such as fires, earthquakes, or severe weather events, and rehearse how to react and communicate effectively in each situation.

Account for Special Needs: Consider any special needs or requirements of family members, such as medical conditions, disabilities, or language barriers, when developing your family communication plan. Ensure that everyone's needs are addressed, and accommodate individual preferences and abilities to facilitate effective communication and coordination.

By implementing a family communication plan and practicing it regularly, you can improve your household's preparedness, cohesion, and resilience in the face of emergencies, enabling everyone to stay informed, connected, and safe during times of crisis.

Long-Term Survival Strategies

In certain emergency scenarios, such as prolonged power outages, infrastructure failures, or societal collapse, sustaining communication over extended periods becomes critical for survival and security. To ensure long-term communication capabilities, consider the following strategies:

Alternative Power Sources: Invest in alternative power sources, such as solar panels, wind turbines, or portable generators, to maintain electrical power for communication devices and equipment during prolonged emergencies. By harnessing renewable energy sources, you can reduce reliance on grid power and ensure continuity of communication even in off-grid environments.

Energy-Efficient Devices: Use energy-efficient communication devices and equipment that consume minimal power and maximize battery life, enabling prolonged operation on limited resources. Select low-power consumption radios, LED flashlights, and electronic devices with efficient power management features to optimize energy usage and extend battery runtime.

Communication Redundancy: Establish communication redundancy by diversifying your communication methods and devices to mitigate the risk of single points of failure. Incorporate multiple communication channels, such as radio frequencies, satellite networks, and internet-based platforms, into your communication strategy to maintain connectivity and resilience across diverse environments and scenarios.

Community Networks: Build and participate in community-based communication networks, such as amateur radio clubs, neighborhood watch groups, or prepper communities, to establish local communication infrastructure and support networks for mutual assistance and collaboration. By pooling resources, sharing knowledge, and coordinating activities, community networks can enhance resilience, self-reliance, and security in long-term survival scenarios.

Information Security: Prioritize information security and operational security (OPSEC) to protect sensitive communication and intelligence from interception, surveillance, or compromise by hostile actors. Use encryption, authentication, and privacy-enhancing technologies to secure communication channels and data transmissions, and practice discretion and caution when sharing sensitive information or operational details.

By implementing these long-term survival strategies and incorporating them into your emergency preparedness plans, you can ensure sustained communication capabilities, resilience, and security in the face of extended emergencies and uncertain conditions. By staying informed, connected, and adaptable, you can navigate challenges, overcome obstacles, and thrive in even the most adverse circumstances.

Building a comprehensive prepper's kit, creating a family communication plan, and implementing long-term survival strategies are essential components of emergency preparedness and resilience. By prioritizing communication readiness, coordination, and adaptation, you can enhance your ability to respond effectively to emergencies, protect your loved ones, and safeguard your well-being in times of crisis. Through proactive planning, preparation, and practice, you can empower yourself and your community to weather any storm and emerge stronger on the other side.

CHAPTER 13: Troubleshooting and Maintenance

Encountering technical issues and equipment malfunctions is not uncommon. Therefore, knowing how to troubleshoot common problems, perform routine maintenance, and handle repairs is essential for ensuring the reliability and longevity of your communication equipment. This chapter provides comprehensive guidance on troubleshooting and maintenance practices to keep your radio in top shape and address issues promptly and effectively.

Common Issues and Solutions

When using radio communication equipment, various issues may arise, ranging from signal interference and poor reception to device malfunction and user error. By familiarizing yourself with common problems and their solutions, you can diagnose issues quickly and implement corrective measures to restore functionality. Some common issues and their solutions include:

Signal Interference: If you encounter signal interference or static noise while using your radio, try adjusting the antenna position, changing frequencies, or relocating to a different location to minimize interference from nearby electronic devices or environmental factors.

Poor Reception: Poor reception can be caused by antenna misalignment, signal obstructions, or weak signal strength. To improve reception, ensure that the antenna is properly installed and oriented for optimal signal reception, and consider using an external antenna or signal amplifier to boost signal strength.

Battery Drain: Excessive battery drain can result from prolonged use, high transmit power settings, or battery aging. To conserve battery life, reduce transmit power levels, turn off unnecessary features, and carry spare batteries or portable chargers for extended operations.

Programming Errors: Programming errors, such as incorrect frequency settings or channel configurations, can lead to communication issues and operational inefficiencies. Double-check programming parameters, consult user manuals or programming guides, and perform factory resets if necessary to correct programming errors and restore proper functionality.

Device Malfunction: If your radio experiences device malfunction or hardware failure, such as button malfunction or display issues, consult the user manual for troubleshooting guidance, contact technical support for assistance, or consider professional repair services for more complex issues.

By systematically diagnosing and addressing common issues as they arise, you can minimize downtime, optimize performance, and ensure reliable communication in various operational environments and scenarios.

Routine Maintenance

Routine maintenance is essential for keeping your radio equipment in top condition and preventing premature wear and tear. By following a regular maintenance schedule and performing preventive maintenance tasks, you can prolong the lifespan of your equipment and minimize the risk of unexpected failures.
Some routine maintenance tasks to consider include:

Cleaning and Inspection: Regularly clean and inspect your radio equipment for dirt, dust, moisture, and debris that may accumulate over time and affect performance. Use compressed air, soft brushes, and cleaning solutions to remove contaminants from external surfaces, connectors, and ventilation ports, and inspect internal components for signs of damage or wear.

Battery Care: Proper battery care is crucial for maintaining battery performance and longevity. Follow manufacturer recommendations for charging, discharging, and storing batteries, avoid overcharging or deep discharging batteries, and periodically test battery capacity and voltage levels to ensure optimal performance.

Antenna Maintenance: Inspect and maintain your antenna system regularly to ensure proper alignment, integrity, and performance. Check antenna connections, cables, and mounts for corrosion, damage, or loose connections, and clean antenna elements and insulators to remove dirt, grime, or oxidation that may degrade signal transmission.

Software Updates: Stay up-to-date with firmware and software updates for your radio equipment to access new features, performance enhancements, and bug fixes released by manufacturers. Check manufacturer websites, forums, or software utilities for update notifications and follow instructions to install updates safely and securely.

By incorporating routine maintenance tasks into your operational practices and adhering to manufacturer guidelines, you can preserve the reliability, functionality, and value of your radio equipment over time, ensuring optimal performance and longevity.

Repair and Spare Parts

Despite proactive maintenance efforts, radio equipment may occasionally require repair or replacement of components due to wear and tear, damage, or malfunction. By having a supply of spare parts and basic repair tools on hand, you can address minor issues promptly and perform field repairs as needed.

Some repair and spare parts considerations include:

Spare Batteries and Chargers: Maintain a supply of spare batteries, chargers, and power accessories to ensure uninterrupted power supply and operational readiness during extended deployments or emergencies. Carry spare batteries in various capacities and chemistries to accommodate different power requirements and usage scenarios.

Replacement Antennas and Accessories: Keep spare antennas, connectors, cables, and mounting hardware in your toolkit to address antenna damage or failure and maintain signal integrity and performance. Choose antennas and accessories compatible with your radio equipment and operational requirements to ensure seamless integration and functionality.

Basic Repair Tools: Equip yourself with basic repair tools and equipment, such as screwdrivers, pliers, soldering irons, and multimeters, to perform minor repairs, adjustments, or modifications to your radio equipment in the field. Familiarize yourself with equipment schematics, technical manuals, and repair procedures to troubleshoot and diagnose issues effectively and safely.

Professional Repair Services: For complex or specialized repairs beyond your expertise or capabilities, consider seeking professional repair services from authorized service centers or certified technicians. Contact manufacturer support channels, authorized dealers, or online repair networks for assistance with warranty repairs, component replacements, or advanced troubleshooting.

By maintaining an inventory of spare parts, tools, and resources, and developing proficiency in basic repair and maintenance techniques, you can address equipment issues promptly, minimize downtime, and extend the operational life of your radio equipment.

Troubleshooting and maintenance are essential aspects of radio communication management, ensuring the reliability, performance, and longevity of your equipment in various operational environments and scenarios. By mastering troubleshooting techniques, performing routine maintenance tasks, and maintaining a supply of spare parts and repair tools, you can optimize the functionality, resilience, and value of your radio equipment, empowering you to communicate effectively and adapt to changing circumstances with confidence and reliability. Through proactive maintenance practices and preparedness measures, you can overcome challenges, mitigate risks, and maximize the utility of your radio equipment, enhancing your capabilities and readiness for mission-critical operations and emergency response efforts.

Chapter 14: Legal and Ethical Considerations

Adherence to legal and ethical principles is paramount to ensure responsible and lawful use of radio equipment. This chapter explores the licensing requirements for operating radios, the ethical considerations of radio communication, and the importance of privacy and security in safeguarding communications.

Licensing Requirements

Operating certain types of radio equipment, particularly those with higher transmit power or operating on specific frequency bands, often requires obtaining a license from regulatory authorities. One of the most well-known licenses in the radio communication community is the Amateur Radio License, commonly referred to as the HAM license. To obtain a HAM license, individuals must pass an examination administered by authorized organizations, demonstrating proficiency in radio theory, regulations, and operating procedures. Licensing ensures that operators possess the necessary knowledge and skills to operate radios safely, efficiently, and in compliance with regulatory requirements. Additionally, licensing provides a framework for accountability and responsibility, promoting responsible behavior and adherence to established standards within the radio communication community.

Ethical Use of Radios

Ethical considerations play a crucial role in radio communication, influencing how operators interact with each other, communicate with third parties, and use radio equipment in various contexts. Ethical use of radios encompasses principles such as honesty, integrity, respect for privacy, and consideration for others. Operators are expected to conduct themselves in a manner that upholds the dignity and reputation of the radio communication community, refraining from engaging in activities that may cause harm, disruption, or offense to others.

Ethical communication practices include:

Courtesy and Respect: Treat fellow operators with courtesy, respect, and professionalism, refraining from engaging in disrespectful or offensive behavior, such as using derogatory language, making personal attacks, or disrupting conversations.

Compliance with Regulations: Adhere to regulatory requirements and operating procedures governing radio communication, ensuring compliance with frequency allocations, power limits, emission standards, and other legal obligations to avoid interference, sanctions, or penalties.

Transparency and Honesty: Be transparent and honest in your communications, providing accurate information, identifying yourself by call sign or handle, and avoiding misrepresentation, deception, or dishonesty in interactions with other operators.

Responsible Operation: Exercise care and responsibility when operating radios, avoiding reckless or careless behavior that may endanger yourself or others, cause interference, or violate operational protocols.

Community Engagement: Participate in community activities, events, and initiatives that promote collaboration, education, and outreach within the radio communication community, contributing to the advancement and welfare of the hobby or profession.

By adhering to ethical principles and norms, operators can foster a culture of trust, cooperation, and mutual respect within the radio communication community, enhancing the overall quality and integrity of communication practices.

Privacy and Security

Privacy and security are fundamental aspects of radio communication, especially in an era of increasing surveillance, data breaches, and cyber threats. Protecting the confidentiality, integrity, and authenticity of communications is essential for preserving individual rights, ensuring operational security, and safeguarding sensitive information from unauthorized access or disclosure. Measures to enhance privacy and security in radio communication include:

Encryption and Authentication: Use encryption and authentication mechanisms to secure communication channels and data transmissions, preventing eavesdropping, tampering, or interception by unauthorized parties. Employ encryption protocols, secure key exchange methods, and digital signatures to protect sensitive information and maintain communication confidentiality and integrity.

Operational Security (OPSEC): Implement operational security practices to minimize the risk of information compromise, surveillance, or exploitation by adversaries. Practice discretion, avoid disclosing sensitive information over insecure channels, and adopt measures to protect communication equipment, locations, and operational activities from detection or surveillance.

Frequency Hopping and Spread Spectrum Techniques: Employ frequency hopping and spread spectrum techniques to mitigate the risk of signal interception or jamming by adversaries, enhancing the resilience and security of communication systems against electronic warfare threats and hostile interference.

Physical Security: Protect radio equipment and infrastructure from physical threats, theft, or tampering by securing access points, employing locks, alarms, or surveillance cameras, and storing equipment in secure locations to prevent unauthorized access or compromise.

Regulatory Compliance: Comply with regulatory requirements and legal obligations governing privacy, data protection, and electronic communications, ensuring adherence to applicable laws, regulations, and industry standards to protect individual rights and privacy interests.

By integrating privacy and security measures into radio communication practices, operators can mitigate risks, safeguard sensitive information, and preserve the confidentiality and integrity of communications in diverse operational environments and scenarios.

Legal and ethical considerations are essential aspects of radio communication management, guiding operators in the responsible and lawful use of radio equipment. By obtaining appropriate licenses, adhering to ethical principles, and prioritizing privacy and security, operators can promote integrity, accountability, and professionalism within the radio communication community, fostering a culture of trust, respect, and cooperation. Through education, awareness, and compliance with regulatory requirements, operators can navigate legal and ethical challenges effectively, ensuring the continued viability and sustainability of radio communication as a valuable tool for communication, collaboration, and community engagement.

CONCLUSION

In this comprehensive guide to Baofeng radios and emergency communication strategies, we have explored a wide range of topics essential for anyone interested in preparedness, survival, and effective communication during emergencies. From understanding the basics of Baofeng radios to developing comprehensive communication plans and addressing legal and ethical considerations, each chapter has provided valuable insights and practical guidance to help readers navigate the complexities of emergency communication.

Summary of Key Points

Throughout the guide, we have emphasized the importance of being prepared and equipped with the knowledge, skills, and resources necessary to communicate effectively in times of crisis. Key points covered include:
- **Understanding Baofeng radios:** We have delved into the history, evolution, and basic functions of Baofeng radios,

highlighting their versatility and applicability in emergency scenarios.

- **Getting started with Baofeng radios:** From purchasing your first radio to unboxing, setup, and basic operation, we have provided step-by-step guidance to help beginners get acquainted with their equipment.

- **Essential accessories:** We have explored the importance of antennas, batteries, headsets, and other accessories in enhancing communication clarity, range, and reliability.

- **Basic radio communication principles:** Understanding radio waves, frequency bands, and proper radio etiquette is crucial for effective communication and interoperability.

- **Programming Baofeng radios:** Whether manually programming or using software tools like CHIRP, mastering the programming process is essential for customizing channels and frequencies to meet specific communication needs.

- **Emergency frequencies and networks:** We have discussed the significance of monitoring emergency channels, joining local radio networks, and leveraging repeaters to stay informed and connected during emergencies.

- **Advanced communication techniques**: Signal boosting, encryption, and repeater use are among the advanced techniques explored to enhance communication capabilities and security.

- **Guerrilla communication strategies**: In challenging environments or hostile conditions, improvising antennas, employing stealth techniques, and setting up base stations are essential for maintaining communication resilience.

- **Prepping essentials:** Building a comprehensive prepper's kit, creating a family communication plan, and implementing long-term survival strategies are critical for ensuring readiness and resilience in the face of emergencies.

- **Troubleshooting and maintenance:** Knowing how to troubleshoot common issues, perform routine maintenance, and handle repairs is essential for maintaining the reliability and functionality of radio equipment.

- **Legal and ethical considerations**: Adhering to licensing requirements, ethical communication practices, and privacy and security measures is vital for responsible and lawful use of radio equipment.

Building a Communication Plan

One of the most important takeaways from this guide is the importance of developing a comprehensive communication plan tailored to your specific needs and circumstances.

A communication plan should encompass various elements, including:
- Identifying communication goals and objectives.
- Assessing communication needs and requirements.
- Selecting appropriate communication methods and devices.
- Establishing communication protocols and procedures.
- Designating emergency contacts and meeting points.
- Conducting regular drills and exercises to practice communication strategies.

By building a communication plan, individuals and organizations can enhance their preparedness, coordination, and resilience, ensuring effective communication before, during, and after emergencies.

Final Thoughts

As we conclude this guide, it is essential to emphasize the importance of being prepared and proactive in managing communication challenges and emergencies. Whether facing natural disasters, man-made crises, or SHTF scenarios, having reliable communication capabilities can mean the difference between safety and peril, survival and adversity.

Staying informed, practicing communication skills, and staying abreast of technological advancements are crucial for adapting to evolving threats and changing circumstances. By embracing a mindset of preparedness, resourcefulness, and community collaboration, individuals and communities can overcome challenges, mitigate risks, and thrive in even the most adverse conditions.

Encouragement to Practice and Stay Informed

We encourage readers to continue practicing their communication skills, staying informed about emergency preparedness best practices, and actively participating in community-based initiatives and networks. By sharing knowledge, experiences, and resources, we can strengthen our collective resilience and empower each other to face whatever challenges the future may bring.

Remember, preparedness is not a destination but a journey, and every step taken towards readiness brings us closer to a safer, more secure future. Let us remain vigilant, adaptable, and committed to the principles of preparedness, resilience, and community solidarity, ensuring that we are always ready to communicate, collaborate, and overcome adversity together.

Thank you for embarking on this journey with us, and may you stay safe, informed, and connected in all your endeavors.

www.ingramcontent.com/pod-product-compliance
Lightning Source LLC
Chambersburg PA
CBHW050059230526
45470CB00004B/1603